M000094285

BLACK AND GREEN

BLACK AND GREEN

AFRO-COLOMBIANS, DEVELOPMENT, AND NATURE IN THE PACIFIC LOWLANDS

Kiran Asher

Duke University Press
Durham and London
2009

© 2009 Duke University Press

All rights reserved

Printed in the United States of America on acid-free paper ∞

Designed by C. H. Westmoreland

Typeset in Galliard with The Sans display by Achorn International

Library of Congress Cataloging-in-Publication Data appear on the last printed page of this book.

TO MY MOTHER,

THE INDEFATIGABLE

VEENA VITHALDAS ASHER

CONTENTS

ACKNOWLEDGMENTS

This project was born when, in a fit of truancy, my friend Heidi Rubio and I abandoned our yoga practice to pore over maps of the Chocó and the new constitution of Colombia. Since that spring morning in 1993, innumerable friends, family members, and colleagues have encouraged and supported me through the long years of research and writing that have led up to this book. I send thanks for and to all of them for sharing my trials and pleasures.

I owe so much to the black communities in Colombia whose courage, struggles, and passion for life inspired this project and keep it alive. The book rests on the time and concerns of black activists, especially those of the PCN and women's organizations. My heartfelt thanks to Carlos Rosero, Libia Grueso, Víctor Guevara, Yellen Aguilar, Leyla Arroyo, and Edwin Murillo, among others in Buenaventura; to Carlos Francisco Ocoró in Guapi; to Hernán Cortés, Julio César Montaño, Martha Arboleda, and José Arismendi in Tumaco; and to Zulia Mena, Saturnino Moreno, and other members of the ACIA and OBAPO in Quibdó. During my fieldwork, the people of the Pacific opened up their homes and hearts to me. Thanks go to Don Miguel Santo and Doña Joba, Doña Naty and Don Poty Urrutia, Doña Esperanza and Don Felipe García in Calle Larga, Río Anchicaya; to Buenaventura Caicedo in Sabaletas; and to Daniel Cristóbal Obregon in El Valle, Chocó. My warmest thanks to Cipriana Diuza Montaño, Sylveria Rodríguez, and other CoopMujeres members, and to Teófila Betancourt in Guapi; Patricia Moreno, Mercedes Segura, Dora Alonso, and Myrna Rosa Rodríguez of Fundemujer in Buenaventura; Nimia Teresa Cuesta, Rosemira Valencia, and Adriana Parra in Quibdó. My gratitude to Jeannette Rojas in Cali for her help and solidarity in connecting me with her networks in the Pacific, and to

Gloria Velasco and Gabriella Castellanos of the Centro de Estudios de Mujer, Genero y Sociedad at the Universidad del Valle for invaluable conversations and for providing access to the center's library.

While it is impossible to name everyone who contributed to this project, I must mention a few more by name: Jaime Rivas, Daniel Riberos, Eduardo Martínez, Pablo Leal, Jorge Ceballos, Cristina Escobar, María Clara Llano, and Betty Ruth Lozano in Cali; Lavinia Fiori, Mónica Restrepo, Claudia Mejía, Mary Lucía Hurtado, Robin Hissong, and in Bogotá. I also acknowledge the help I received from Álvaro Pedrosa, María de Restrepo, Elías Sevilla, and Fernando Urrea at the Universidad del Valle; Nina de Friedemann and Jaime Arocha at the Universidad Nacional; and Eduardo Restrepo, David López, Aurora Sabogal, and Óscar Alzate at Bosques del Guandal. Very special thanks to Claudia Leal, Santiago Muñoz, and Andrés Guhl for helping to make the maps for this book. Comments and insights about Afro-Colombian struggles from Peter Wade, Joanne Rappaport, Arturo Escobar, Claudia Mosquera, and Ann Farnsworth-Alvear have been invaluable for my research. Thanks also to Steven Sanderson and Debbie Pacini who were then at the University of Florida and to Jean Brown, who will take special pride in the book.

The early phase of research for this book (1993–95) was funded by the Tropical Conservation and Development Program at the University of Florida, the Inter-American Foundation and the Fundación para la Educación Superior (FES). In 1995, the Department of Sociology of the Universidad del Valle hosted me as an invited researcher; Fundación Habla/Scribe and Fundación Herencia Verde provided logistical and administrative support. Follow-up work in Colombia was possible through small grants from Bates College (1999) and Clark University (2005, 2007). A Rockefeller postdoctoral fellowship (2001–2002) at the Institute for Research on Women, Rutgers University, and a Five College Women's Studies Research Center fellowship (spring 2006) provided invaluable time for writing. While there, the conversation and camaraderie of the other fellows was key. I especially thank Mariella Bacigalupo, Negar Mottahedeh, and Jasbir Puar. I also thank the editors of the journals *Feminist Review* and *Feminist Studies* for publishing my early insights on black women's struggles. Chapter 4 of this book is a revised and differently contextualized version of that early work.

The intellectual labor of this book was accompanied by much angst and joy. Friends helped lighten the former and heighten the latter. Camila Moreno, Heidi Rubio, Ernesto Ráez-Luna, Luis Miguel Renjifo,

Alberto Gaona, Claro Mina, Beto Valdés, and Juana Camacho are certainly *amigos en las buenas y en las malas*. I also make special mention of Lise Waxer, my housemate in Cali while we were both beginning our doctoral research, who introduced me to salsa and Afro-Latin music (the foci of her research). She died suddenly in 2002, and I regret that I never had a chance to tell her how much I learned to love the music she introduced me to. After a year of Colombian sunshine, I returned to Gainesville in December 1995 during one of the coldest winters in Florida. Charo Lanao and Benjamín Vivas kept joy alive in my life with their love and contagious laughter. Many of the book's arguments were worked out over conversations and meals with Holly Hanson, Marcia Good Maust, Kathryn Burns, Rebecca Karl, Anindyo Roy, Bettina Ng'weno, and Joel Wainwright. Time with them and with Leslie Hill, Rebecca Herzig, Pat Saunders, and Kate Adams in Maine helped me see the joy of academic and political struggles. Miriam Chion, Dave Bell, Amy Ickowitz, Anne Geller, and my students (especially Diana Ojeda, Allen Gallant, and Sheryl-Ann Simpson) at Clark allowed me to keep my work alive in Massachusetts. I am grateful to Valerie Millholland, Miriam Angress, Tim Elfenbein, and others at Duke University Press for their patience and professionalism in taking the book to print.

Thanks to my families (the Ashers, including Nina Asher and the Santa Cruz Ashers; the Ghelanis; and the Redicks) for their pride and even pleasure in my work. Finally, thanks to my *compañero* Robert Redick for reading countless drafts of this manuscript over the course of a decade, for his incomparable editorial support, and for believing in me during my moments of deepest self-doubt. I alone am responsible for the errors in the book.

ABBREVIATIONS AND ACRONYMS

ACADESAN	Asociación Campesina del Río San Juan (Peasant Association of the San Juan River)
ACAPA	Consejo Comunitario de Comunidades Negras del Río Patía Grande, sus Brazos, y la Ensenada (Black Community Council of the Greater Patia River Basin)
ACIA	Asociación Campesina Integral del Río Atrato (United Peasant Association of the Atrato River)
ACNUR	Alto Comisionado de las Naciones Unidas para los Refugiados
AFRODES	Asociación de Afrocolombianos Desplazados (Association of Displaced Afro-Colombians)
ANUC	Asociación Nacional de Usuarios Campesinos (National Association of Peasant Producers)
AUC	Autodefensas Unidas de Colombia (United Self-Defense Groups of Colombia)
CAVIDA	Comunidades de Autodeterminación, Vida, Dignidad del Cacarica (Cacarica Communities for Self-Determination, Life, and Dignity)
CCAN	Comision Consultiva de Alto Nivel (High-Level Consultative Commission)
CECN	Comisión Especial para las Comunidades Negras (Special Commission for Black Communities)
CODECHOCO	Corporación Autonóma Regional para el Desarrollo del Chocó (Chocó Autonomous Regional Development Corporation)
CODHES	Consultoría para los Derechos Humanos y Desplazamiento (Consultancy on Human Rights and Displacement)

COHA	Council on Hemispheric Affairs
CONPES	Consejo Nacional de Política Económica y Social (National Council for Economic and Social Policy)
CORPONARIÑO	Corporación Autónoma Regional de Nariño (Nariño Autonomous Regional Development Corporation)
CORPURABA	Corporación para el Desarrollo Sostenible del Urabá (Uraba Sustainable Development Corporation)
CRC	Corporación Autónoma Regional del Cauca (Autonomous Regional Development Corporation of Cauca)
CVC	Corporación Autónoma Regional de Valle del Cauca (Autonomous Regional Corporation of Valle del Cauca)
DNP	Departamento Nacional de Planeación (National Planning Department)
ELN	Ejército de Liberación Nacional (Army of National Liberation)
ETI	Entidades Territorials Indígenas (Indigenous Territorial Units)
FARC	Fuerzas Armadas Revolucionarias de Colombia (Revolutionary Armed Forces of Colombia)
FES	Fundación para la Educación Superior (Foundation for Higher Education)
GAD	Gender and development
GEF	Global Environment Facility
ICANH	Instituto Colombiano de Antropología e Historia (Colombian Institute of Anthropology and History)
IGAC	Instituto Geográfico Agustín Codazzi (Agustín Codazzi National Geographic Institute)
IIAP	Instituto de Investigaciones Ambientales del Pacífico John von Neumann (John von Neumann Institute for Pacific Environmental Research)
INCODER	Instituto Colombiano de Desarrollo Rural (Colombian Institute for Rural Development)
INCORA	Instituto Colombiano de Reforma Agraria (Colombian Land and Agrarian Reform Institute)
INDERENA	Instituto de Desarrollo de los Recursos Naturales Renovable (Institute for the Development of Renewable Natural Resources)
OBAPO	Organización de Barrios Populares y Comunidades Negras de Chocó (Organization of People's Neighborhoods and Black Communities of the Chocó)

OCN	Organización de Comunidades Negras (Organization of Black Communities)
OREWA	Organización Regional Emberá–Waunana del Chocó
PCN	Proceso de Comunidades Negras (Process of the Black Communities)
PDCN	Plan de Desarrollo de las Comunidades Negras (Development Plan for Black Communities)
PLADEICOP	Plan de Desarrollo Integral para la Región Pacífico (Integrated Development Plan for the Pacific Coast)
PMRN	Programa de Manejo de Recursos Naturales (Program for the Management of Natural Resources)
PNR	Plan Nacional de Rehabilitación (National Rehabilitation Plan)
RSS	Red de Solidaridad Social (Social Solidarity Network)
UNCED	United Nations Conference on Environment and Development
UNDP	United Nations Development Program
UNHCR	United Nations High Commissioner for Refugees
UNICEF	United Nations Children's Fund
USAID	U.S. Agency for International Development
WID	Women-in-development
WOLA	Washington Office on Latin America

Map 1. Map of Colombia with major national cities and location of the Pacific. Prepared by Claudia Leal and Santiago Muñoz; modified by Tracy Ellen Smith.

Map 2. Map of the Colombian Pacific region with regional center and major rivers. Prepared by Claudia Leal and Santiago Muñoz; modified by Tracy Ellen Smith.

INTRODUCTION

Black Social Movements and Development

in the Making

Capitalist development has not been unmade by Third World social movements; rather, both development and such movements are being restructured during the current neoliberal, free-market phase of capitalism. This is not to say that development in its globalizing guise will (or can) fulfill its promises of promoting economic growth, meeting basic human needs, managing natural resources sustainably, and conserving the environment. But neither are alternatives proposed by new social movements of marginalized communities of women, workers, peasants, and ethnic groups imminent. However, viewing the development project and social movements in oppositional terms obscures the contradictory, complex, and contingent ways in which they are intertwined. This book moves beyond the notion that development is a hegemonic, homogenizing force of Western rationality to show how its discourses and practices structure, and are structured by, their reception in particular locations at specific conjunctures. Specifically, it is an ethnographic study of the processes of black social movements as they emerged from and articulated with political economic changes in the Pacific lowlands of Colombia in the 1990s. More broadly, the work theorizes these dynamics beyond the many binaries—tradition versus modernity, progress versus underdevelopment, exploitation versus resistance, local versus global, theory versus practice, identity versus strategy—that plague and limit thinking about Third World development and social movements.

That development and cultural processes unfold together within specific conjunctures of space, time, and power relations, and that these interactions have contingent outcomes was evident from the start of my research. On May 26, 1993, a series of coincidences found me at a session of the Colombian Congress, assembled to discuss a proposed law, stipulated but not described in the new Colombian Constitution. The constitution, ratified but two years before, included Transitory Article (Artículo Transitorio) 55, or AT 55, requiring the timely adoption of a law recognizing

> the collective property rights of black communities that have inhabited the empty lands (*tierras baldías*) in the rural riparian zones of the Pacific coast, in accordance with their traditional production practices [and establishing] mechanisms for the protection of the cultural identity and rights of these communities, and to promote their economic and social development. (Republic of Colombia 1991)[1]

The "empty lands" in question are Colombia's Pacific lowlands, geographically the largest area of black culture in the country. Part of the Chocó biogeographic region, the lowlands extend 1,300 kilometers from southern Panama to northern Ecuador along the Pacific coast. A global biodiversity "hot spot," the region is home to a variety of ecosystems (coral reefs, mangroves, rock and sandy beaches, coastal forests, high- and lowland tropical moist forests) and myriad plant and animal species, many endemic.[2] In the early 1990s, most of this region was yet to be overrun by drug cultivators and traffickers, guerrillas or paramilitary forces. It was better known as a supplier of natural resources: timber, gold, platinum, silver, oil, and natural gas.

The granting of collective land titles to Afro-Colombian communities (who make up 90 percent of the region's population) seemed at odds with global, Latin American, and, indeed, national trends, all emphatically in the direction of privatization. But the framers of the 1991 Constitution had many agendas. Significant among them were neoliberal economic reforms targeting marginal areas such as the Pacific for natural-resource extraction and modernization. At the capitol building that morning, it became evident that a law granting land rights to black communities was by no means a fait accompli. This was not only because the Colombian state had other economic interests in the Pacific, but also because of diverging opinions regarding the parameters of any such law.

I went to Congress with a group of activists and a sociologist from the Special Commission for Black Communities (Comisión Especial para

las Comunidades Negras; CECN), appointed by the Colombian government to draft the new law. When our delegation, made up predominantly of brown and black people (including myself), arrived at the capitol building in downtown Bogotá, we were denied entry. It was only
after the sociologist (the lightest-skinned among us and the only one
with a state-issued identity card) intervened that the guard at the door
allowed us into the building. We were neither the first nor the last to
arrive. By 10 A.M., a large crowd had assembled outside the Salon Boy-
aca, where the session was convened. Among the crowd were numerous black activists and *comisionados* (commissioners) of the Congress-
appointed Special Commission.

At 11 A.M., the group was abruptly informed that the session would
not take place. Angry muttering commenced: this was not the first time
that a session scheduled to discuss AT 55 had been inexplicably canceled.
To the visitors it was simply another instance of the Congress's foot
dragging over the proposed black law. Given that it was a few months
before the deadline to pass a law based on AT 55, such stalling had serious overtones for black rights. Few of the activists were inclined to
walk away. After consultation with activists and comisionados, Piedad
Córdoba, a black senator from the Liberal Party of the Department of
Antioquia, browbeat some dozen congressmen into holding an informal
meeting inside a large hall in the capitol building.

The congressmen listened restlessly as activists and comisionados
made loud and impassioned calls for the ratification of AT 55. But the
black delegates did not speak with one voice. Three broad factions emphasized distinct aspects of Afro-Colombian rights. First, there was a
group in favor of a law for black communities, but not as outlined in AT
55. This group—predominantly urban middle-class Afro-Colombians,
many of whom seemed to have ties to the country's main political parties—considered the notion of collective land titles for black communities a regressive one. A dark-suited member of this contingent argued
that blacks were a marginalized minority in Colombia and that special
legal measures were needed to promote social, political, and economic
equality of all black communities, not just those living in the Pacific
region.

The second faction consisted of delegates from the Department of
Chocó—a black-majority department on the northwest Pacific coast—
for whom the issue of land titles for black communities was a matter of
urgency. In this group stood Saturnino Moreno, a wiry old *campesino*
from the Atrato River. Bowler hat in hand, Moreno described how elite

black politicians in the Chocoan capital, Quibdó, had granted extensive logging concessions over tracts of forests to private firms. The resulting commercial logging was destroying the natural-resource base and displacing communities. For Moreno, collective titles to forest lands would enable local communities to secure their livelihoods and make sound decisions regarding the sustainable use and management of the region's natural resources.

Zulia Mena, a young woman who worked with squatters and migrants in the shantytowns of Quibdó, supported Moreno's position, arguing that the state must guarantee the land rights of black communities. Invoking language in the constitution that declares Colombia a multiethnic and pluricultural nation, she added that a black law should also protect cultural differences *among* Afro-Colombians. A year later, Mena would be elected to Congress.

Another supporter of a law for black communities was Nina de Friedemann, a well-known anthropologist and scholar of Afro-Colombian history and culture. De Friedemann spoke of Afro-Colombians and their contributions to the nation's history, economy, and culture and how they were "invisibilized" because of ethnic discrimination and the nation's racist political ideology. She ended by urging the government to "right historical wrongs" and legally recognize Afro-Colombian ethnic and territorial rights.

The third group, delegates from the Organization of Black Communities (Organización de Comunidades Negras; OCN), spoke the longest. The OCN spokesman began his long oration by agreeing with his *compañeros* that a black law should guarantee the rights of all Afro-Colombians, not just those living in the rural areas of the Pacific. Noting that AT 55 had been included in the new constitution under duress, he accused state officials of having *mala fe* (bad intentions) vis-à-vis black rights. He went on to denounce various state-sponsored plans for the region—roads, dams, oil pipelines, commercial forestry, aquaculture, agriculture, mining—and argued that such capital-intensive development in the name of modernization was destroying black culture and the biodiverse natural world.

When he concluded, another OCN delegate outlined a proposal for an "autonomous collective ethnic territory" in the Pacific region, where diverse black and indigenous communities could come together and negotiate their own culturally appropriate and ecologically sustainable visions of development. He noted that Afro-Colombians' struggles were not just about equality and collective land rights; they were, rather, about the right to be (*ser*) black, to celebrate their cultural identity, and to live in accordance with tradition. The position of the OCN, it appeared, was that Afro-Colombian rights meant having territorial and administrative control over the Pacific region together with the other ethnic groups resident there. The delegate observed that black communities had lived in the Pacific region for over five hundred years and developed economic subsistence practices that sustained their culture and environment. It was in this "homeland" that black communities and their indigenous neighbors had nurtured social relations based on mutual respect and peaceful coexistence. He reiterated that protecting black culture in Colombia depended on more than equality and land rights. It required collective, autonomous territory. I later learned that the OCN was a loose coalition of activists and intellectuals from three southern coastal Pacific states (Valle del Cauca, Cauca, and Nariño), as well as individuals and interest groups from Bogotá and the departments of Atlántico and Antioquia.

By this point, the congressmen were visibly impatient. One declared that there was no question of discriminating against Afro-Colombians, that the constitution protected the rights of all citizens, and that the socioeconomic development of black communities in the Pacific region was a state priority. The meeting adjourned inconclusively.

But that summer, across the country, Afro-Colombian sectors came together to mobilize black communities and to discuss and negotiate draft proposals for a black law. They formed alliances with indigenous groups and others sympathetic to their struggles and lobbied intensively for the legal recognition of black rights. Finally, on August 27, 1993, President César Gaviria approved "Ley 70, La Ley de las Comunidades Negras (Law 70, the Law of Black Communities)."[3]

Law 70's eight chapters and sixty-eight articles focus on three main issues: ethnic and cultural rights, collective land ownership, and socioeconomic development. In its own words, the law

> recognizes that black communities occupying the *tierras baldías* (empty lands) of the rural, riverine zones of the Pacific river basins and using traditional practices of production have the right to collective land titles in accordance with the following articles. The law also proposes to establish mechanisms to protect the cultural identity and the rights of the black communities of Colombia as an ethnic group, and to promote their economic and social development to guarantee that these communities obtain real conditions of equal opportunity vis-à-vis the rest of Colombian society.

In accordance with Paragraph 1 of Transitory Article 55 of the Constitution, this Law will also apply to rural, riverine *baldía* zones in other parts of the country occupied by black communities who use traditional practices of production and satisfy the requisites established in this law. (Article 1, Law 70 of 1993)

Law 70 does expand the terms of AT 55 to recognize the rights of all Afro-Colombians, but in restricted terms. These rights are modeled on the perceived realities of black communities of the Pacific and emphasize collective property titles for groups living in rural areas and engaged in subsistence production. However, black communities were not, as the OCN proposed, granted autonomous control over the Pacific region.

Afro-Colombian rights were only one part of the state's agenda in the Pacific. In a country with complex geography where regions are often physically and politically disconnected from each other, the Chocó was long relegated to the periphery of post-World War Two national development. With the arrival of neoliberal globalization in the late 1980s, however, state officials and development experts held that it was strategically and economically imperative to modernize hitherto isolated regional economies (such as in the Pacific and Amazon areas) and to integrate their "backward" inhabitants with the rest of the country. The Pacific littoral became a particular target of economic interventions because of its extensive reserves of natural resources and immense genetic diversity. Even as the parameters of black rights were being debated, state planners were engaged in talks with multilateral donors to fund a large-scale economic initiative. Called Plan Pacífico, its explicit aim was to develop the region's natural resources and stimulate economic growth.

The biodiverse Chocó was also fast becoming a key target of national and international environmental conservation. In 1992, the Colombian government launched a five-year biodiversity conservation program called Proyecto BioPacífico. With a mandate to devise mechanisms for the protection and sustainable use of regional biodiversity, Proyecto BioPacífico became linked to the economic aims of Plan Pacífico. As awareness of Plan Pacífico and Proyecto BioPacífico spread, ethnic groups drew on new global discourses of "rights-based development" and "community-based conservation" and began to pressure the state to make good on its promises to recognize their rights. In response to such pressure, the collective titling of ethnic lands, local participation, and preservation of traditional knowledge of natural resource management became subsidiary goals of economic development and environmental conservation mandates.

It was a time of significant victories, then. Yet the passing of Law 70 and provisos to address black rights within the development agenda did not unite Afro-Colombians under a single organizational umbrella. On the contrary, in the aftermath of the law's passage, differences became further entrenched along the three lines I witnessed at the Congress hearing. Mainstream black politicians capitalized on the national momentum around Afro-Colombian rights to seek support of the black electorate for their candidacy to political offices at national and regional levels. Chocoan groups asserted the primacy of ethnic and territorial rights to break into the clientelist stronghold of department politics dominated by black elites. The OCN, renamed the Process of the Black Communities (Proceso de Comunidades Negras; PCN), called for an autonomous black "ethno-cultural" movement and culturally appropriate alternatives to the economic development agenda.

Alongside these factions came a burgeoning of wider activism, as increasing numbers of black groups—peasants, urban squatters, women, youth groups—turned to Law 70 to stake claims on the state or draw support for specific local struggles, which were increasingly couched in ethnic terms. In sum, post-Law 70 black social movements and the political-economic dynamics of development in the region were deeply intertwined and remained fraught with tensions and contradictions.

This book explores these tensions and contradictions. It argues that political-economic processes and struggles for social change shape each other in uneven and paradoxical ways. Starting with a discussion of how and why black rights were included in the new constitution, the book focuses on the claims made in the name of culture, nature, and development by Afro-Colombian groups and the Colombian state. Through an ethnographic study of black organizing in the Pacific in the 1990s, I show how local struggles and cultural politics are constituted differentially, unequally, and discursively by and against modernizing or globalizing interventions.

As macro-development unfolded in the Pacific region (as elsewhere), local and global processes became increasingly intertwined. This linkage implied neither the disappearance of the state nor a homogenization of social and cultural relations, however. The result was rather an explosion of movements contesting state power while simultaneously seeking state recognition. Black communities linked their demands for ethnic and territorial rights to the latest phase of capital intrusion and state formation in the Pacific that had acquired "environmentally sustainable" and "culturally sensitive" hues. The state's economic growth plans and ethnic territorial claims emphasized "biodiversity conservation" and "sustainable natural-resource management." However, both state plans and black rights become objects of contestation as local communities, state entities, and nongovernmental organizations (NGOs) of various stripes attempted to give meaning and shape to the new interventions in accordance with their own understanding and interests. As noted earlier, contradictions and differences also abounded within black movements themselves. While never merely struggling for inclusion, black communities were neither outside the discourses of development nor content to remain on its margins. Rather, they were constituted as subjects of prevailing development and ethno-cultural discourses even as they attempted to disrupt them. By exploring how social movements and development processes emerge in relation to each other, and taking their differences and contradictions seriously, this work conceptualizes local communities beyond "victims of development" or "defenders of tradition."

Development, the Environment, and Ethnic Rights in the Pacific Lowlands

Development (and its close relative, modernity) has a long history, with links to eighteenth-century Enlightenment ideals of progress through reason. In the nineteenth century and early twentieth century, they unfolded in close conjunction with the industrial and scientific revolutions centered in Europe. The concomitant capitalist, colonial, and imperial expansions rapidly and drastically changed the conditions of existence all over the globe. These transformations also led to intense political, economic, social, and cultural turmoil. They did not, however, solve the ills of society.

In the post-World War Two period, on the heels of formal decolonization, the idea of promoting political and economic modernity through development in the newly independent Third World nations gained particular salience. This phase of development, like its colonial and imperial precursors, was complex and contested and meant different things to different people—inspiring scholars and practitioners, seducing politicians and revolutionaries alike. In the early decades of the Cold War, it acquired the technocratic garb of "modernization theory" and "developmentalism," forgot its historical roots in Western Enlightenment, and spoke of its anticommunist mandate largely sotto voce. Throughout the 1960s and 1970s, developmentalism's promoters set about using their technical knowledge to help poor people in the newly independent (African and Asian) and as-yet-underdeveloped (Latin American) countries "catch up" with the living standards of the developed world. But the path to progress, as well marked and straight as many found it, turned out to be full of roadblocks. Among the numerous problems of underdevelopment were those identified by political scientists (e.g., Samuel Huntington on the inadequate institutionalization of power, leading to political corruption and anarchy), economists (e.g., Walter Rostow on insufficient and inefficient growth), sociologists (e.g., Alex Inkeles on inadequate commitment to modern social relations and identities, including those of capitalist industrial production and secular nationalism), anthropologists (e.g., Clifford Geertz on symbolic and material practices and values of "premodern" cultures and agrarian communities and their linkages with modernity), and gender experts (e.g., Ester Boserup on the various exclusions underdevelopment engendered). It was the undisputed task of states to work in concert with international institutions and agencies (the World Bank, International Monetary Fund, United Nations) to address these problems and get development going.

Critics of a more political persuasion continued to converse with the ghost of Marx (Wolf 1982). These critics saw the struggle against underdevelopment as part of anti-imperialist and nationalist struggles. Development economists and dependency and world-systems theorists offered various structural explanations of underdevelopment, arguing that capitalist production (whether inside the boundaries of a nation-state or global) was premised on either the "development of underdevelopment" or the simultaneous creation of a "core" and a "periphery" (Raúl Prebisch, Fernando Henrique Cardoso and Enzo Faletto, Andre Gunder Frank, Immanuel Wallerstein, Samir Amin). These and other critiques were incorporated into official policy. Development schemes acquired numerous taglines: anti-poverty, basic human needs, rights-based, participatory, sustainable.

In the 1980s and 1990s, both advocates (Rodrik 2000; Williamson 2002) and critics (Apffel Marglin and Marglin 1990; Escobar 1992; Esteva 1987; Rahnema and Bawtree 1997; Sachs 1992; Sheth 1987; Shiva 1988) judged the development project to be in crisis. The former, supported by a consensus in Washington, still believed in bringing modernity and progress to the Third World through growth led by the private sector. The latter denounced development as a subset of Eurocentric modernity and began writing its obituaries.

The fall of the Berlin Wall in 1989 and the end of the Cold War ostensibly signified the triumph of capitalism and democracy. Neo-modernizationists exhumed Marx's ghost, only to bury it deeper. With their liberal political and neoliberal economic bents, they emphasized free trade to promote efficient growth and (minimal) policies to address heretofore marginalized social, cultural, and environmental issues with the aid of global market forces and "good governance." Post-developmentalists underscored the power of "the local" and emphasized the need to uncover non-Western rationalities, ones that might have remained outside development's hegemony and faced foreclosure under neoliberal globalization. Though deploying very different rhetorics, each side called for the retrenchment of the state and the rise of "civil society."[4]

At roughly the same time, many Latin American nations, from Mexico to Chile, were in the process of adopting political and economic reforms following the end of military regimes and debt crises of the 1980s. In this period, over a dozen Latin American countries either substantially modified their constitutions or adopted new ones.[5] Colombia's constitutional self-examination was not exceptional in this regard. It was, however, the product of a specifically Colombian history and trajectory.

Since its independence from Spain, modern Colombia has been characterized by intense and violent social turmoil. Eight civil wars were fought in the nineteenth century alone, including the devastating War of the Thousand Days (1899–1902). The twentieth century was no less troubled, being peppered with recurring stretches of violence related to economic and political exclusion, drug trafficking, and paramilitary and left-wing guerrilla activities. It has been argued that all Colombian conflicts of the twentieth century in some way lead up to or follow from a time of intense violence known as La Violencia, which occurred during the decade 1948–58. The murder of the Liberal populist President Jorge Eliécer Gaitán on April 9, 1948, is considered a catalytic moment of La Violencia, a period whose causes, consequences, numbers of dead and duration are subject to much debate. It is estimated that 200,000–300,000 Colombians, mainly peasants, lost their lives in the sectarian violence in the decade following Gaitán's death. To curb this massacre, in 1958 politicians from the Liberal and Conservative parties entered a power-sharing agreement called the National Front. But as political scientists and historians of Colombia (Bushnell 1993; Bergquist et al. 2001; Palacios 2006; Richani 2002) have discussed at length, the prevailing political oligarchy, with a weak two-party system, failed to address the fundamental social and economic problems facing the vast majority of Colombians or to curb the recurring bouts of violence.[6] This violence accelerated, and the Colombian state faced another crisis of legitimacy after 1974, when the National Front collapsed.

Despite having the dubious distinction of being one of the most violent countries in South America, unlike its Southern cone neighbors, the Colombian state did not adopt bureaucratic-authoritarian models of economic growth and political control. John Sheahan (1987) describes Colombia as a "middle-road market economy" in which economic growth through industrialization and modernization of agriculture has flourished since the 1930s, aided by state intervention, elite-dominated private enterprises, and limited political participation.[7] These economic policies led to fair gains in employment, education, and some reduction in rural poverty and decreased dependence on primary exports (Palacios 2006; Sheahan 1987). Marco Palacios notes that during this period there was an increase in coffee production because of agricultural modernization, the emergence of an agricultural middle sector, and increasing parity between rural and urban wages. However, for the most part Colombia's economic programs were short on social goals, and the benefits of this growth were distributed unequally. Skewed distribution of land and income, poverty, and limited democratic participation were chronic problems in both urban and rural areas and led poor Colombians to rise in repeated protest against the ruling oligarchy. Several guerrilla groups emerged during this period, among them the Revolutionary Armed Forces of Colombia (Fuerzas Armadas Revolucionarias de Colombia; FARC), the Army of National Liberation (Ejército de Liberación Nacional; ELN), the Popular Liberation Army (Ejército Popular de Liberación; EPL), the April 19 Movement (Movimiento 19 de Abril; M-19), and a host of other, smaller groups. These guerrillas, urban squatters, students, unionized factory workers, peasants, and indigenous groups had engaged in various forms of protest against state policies and corruption since the 1970s.

At the end of the 1980s, amid charges of corruption and incompetence, the unrest came to a head. In 1989, a peace accord was signed between President Virgilio Barco and M-19, a democratic alliance of disarmed guerrilla groups. Subsequently, the Colombian state brokered a peace process of sorts and responded to calls for constitutional reform.[8] In 1990, representatives of an elected body—the National Constituent Assembly—wrote and discussed a draft of a new constitution whose aim was to expand national politics, redefine the state's role, and address the persistent problems of underdevelopment and violent conflict. The Constituent Assembly was an unusual council, consisting of representatives from ethnic, religious, and demobilized guerrilla groups, among others. Because of divisions among the traditional political parties, the assembly had unprecedented access to the reform process.

This was not Colombia's first constitutional-reform endeavor, nor was it smooth. But in response to internally induced changes and forces of regional and global geopolitics (promoting economic reforms and democratization yet again), the process went ahead, and a new constitution was ratified on July 4, 1991.[9] The document is an ambitious political charter that aims to expand democratic participation, strengthen civil society, decentralize elite-dominated political administration, and promote economic growth and development. Its principal focus is the creation of a "modern" state ready to play a significant role in national, regional, and global affairs. This is reflected in the constitution's liberal-democratic language and its emphasis on rights, whether political, economic, social, or environmental. A key mechanism specified to ensure access to these rights is political participation along the legislative, executive, and judicial axes of state power. The constitution also mandates reform of the elections process and a reorganization of the territorial and bureaucratic structures of state administration.

Ethnic and cultural issues were not a crucial part of the constitutional-reform agenda initially. But the presence of indigenous representatives in the Constituent Assembly and discussions about the changing nature and meaning of national identity and citizenship led to the inclusion of Article 7 in the new constitution. Under Article 7, the state recognizes and protects the ethnic and cultural diversity of the nation. It was this article that Zulia Mena brought up at the Congress event and that was repeatedly invoked in debates regarding black ethnic rights.

In contrast to this new multiculturalism, the 1991 Constitution's economic reforms were in keeping with earlier state policies and the prevalent neoliberal agenda. The new policies centered on restructuring the national economy to make it more competitive in regional and international markets. With as-yet-undeveloped natural-resource assets (including oil and gas deposits) and proximity to the economically powerful countries of the Asia-Pacific Rim, the Pacific region had important potential for economic growth. It was within this context that Plan Pacífico was drafted as (among other things) an investment guide and blueprint for infrastructure development.

At the same time, regions such as the Amazon and the Pacific emerged into the spotlight of a new era of environmental politics. Concerns over development-related environmental problems (industrial waste, air and water pollution, resource degradation and scarcity, etc.) were added to the global agenda at the United Nations Conference on the Human Environment in 1972. In the 1980s, the concept of "sustainable development" began to link issues of economic development with environmental conservation, most famously in the report published by the World Commission on Environment and Development (1987; see also Redclift 1987). The 1992 United Nations Conference on the Environment and Development in Rio de Janeiro consolidated these linkages, and by the early 1990s, international development entities had absorbed the idea, at least rhetorically, that economic growth and social justice were connected to environmental-ecological concerns (Lele 1991). Influenced in part by these "green" discourses, the new constitution introduced legislation to protect and manage biodiversity (naming it part of the "national patrimony" with "benefits for humankind"). A series of institutions and programs (such as Proyecto BioPacífico) was set up to catalog the country's biodiversity, make plans to conserve it, and develop means to use it profitably, sustainably, and equitably.

During the social and political turmoil of the 1970s and 1980s, black and indigenous communities were among those who protested the state's disregard for their interests and welfare. In rural areas, many of these groups were part of peasant protests for land. Indigenous and black struggles also have long histories, with the indigenous movement achieving particular visibility and prominence in national and international arenas since the 1970s (Dover and Rappaport 1996; Findji 1992; Gros 1991). Prior to the constitutional-reform process, black demands were rarely couched in terms of the recognition of their cultural difference. (Reasons for this are discussed in chapter 1.) But by the early 1990s, both groups were responding to new influences and inspirations. Among these were the uniquely diverse Constituent Assembly (with three indigenous representatives), national debates in favor of multiculturalism, and such international agreements as the International Labor Organization's Accord 169, which demanded the recognition of ethnic and cultural rights of peoples within nations.

Colombia was not the only Latin American country to adopt official multiculturalism. Pro-indigenous policies were also adopted in Argentina, Brazil, Ecuador, Guatemala, Honduras, Mexico, Nicaragua, Paraguay, and Venezuela. Afro-Latin Americans did not fare as well, except in Colombia and Brazil. From Central America (Guatemala, Honduras, Nicaragua) to the central Andes (Ecuador, Bolivia) to Brazil, indigenous movements that had intensified in the 1980s allied themselves with transnational advocacy networks to challenge dominant conceptions of citizenship and to demand territorial autonomy, self-determination, respect for customary laws, and other rights based on reconstructed notions of identity (Brysk 2000; Keck and Sikkink 1998; Van Cott 2000; Warren and Jackson 2002; Yashar 1998). And like other aboriginals in Latin America and beyond, indigenous and black communities made alliances with environmentalists, who saw rural, grassroots communities as "stewards of nature" and advocated their participation in efforts toward sustainable development and environmental conservation. But the appeal to indigeneity, multiculturalism, and environmentalism worked differently for the two ethnic groups.

Since colonial times, indians in Colombia have been perceived as culturally distinct entities.[10] Though at least as often honored in the breach as in the observance, special rights for indigenous Colombians have enjoyed a longstanding conceptual history. Under the 1991 Constitution, indigenous groups were promised expanded control over their communal lands and administrative autonomy in judicial decisions, finances, and development policy. Blacks went into the constitutional-reform process in an ethnic double bind: discriminated against or exoticized because of their "racial difference," but not considered sufficiently distinct from Colombia's *mestizo* (mixed-race) population as indians to merit special legal status. It must be recalled that AT 55's restricted terms for black rights are in part an expression of this ambivalence. Another such expression, evident at the Congress meeting (and explored in the next chapter) was the two-year struggle to make a law of substance out of the constitution's broad language.

Post-Law 70 Black Movements: Contesting Development, Constituting Black Ethnicity

The tensions between "economic and political modernity" and "black ethnic identity and traditional rights" that I had glimpsed during my short trips in 1993 and 1994 were on full display in 1995. On January 20, 1995, the newly established Ministry of the Environment held a public forum (*audencia pública*) at the Universidad del Valle (UniValle) in Cali to discuss the repositioning and replacement of an oil pipeline. The new pipeline was to run from near Cali in the Andean interior of Valle del Cauca Department to the coastal port of Buenaventura via the town of Dagua. This pipeline was part of a larger infrastructure project, one of many conceptualized under Plan Pacífico.

When I arrived at the UniValle auditorium at 11:00 A.M., an engineer from Ecopetrol, the public company responsible for constructing the pipeline, had the podium. For an hour and a half, the engineer spoke about the region's growing domestic and industrial need for energy, the legal and technical aspects of pipeline construction and placement, and the environmental impact assessment that Ecopetrol had commissioned. After this presentation, the official from the Ministry of the Environment presiding over the forum invited questions and comments from the audience. Carlos Rosero and Yellen Aguilar, two dreadlocked PCN leaders (the former an anthropologist trained at the Universidad Nacional; the latter an agronomist), rose from their seats and began to hail the podium with questions about the impact of the proposed pipeline on the lands and lives of black communities. Why had the Ministry of the Environment not informed potentially affected communities in Dagua about the project or the public meeting in a timely manner? What was the difference between the initial project proposed by Ecopetrol and the one under discussion? What mechanisms of participation had the Ministry of the Environment established, or did it plan to establish, to ensure that affected communities could put forward their ideas, objections, and questions? Had socioeconomic studies been conducted on the potential impact of the project? Had local communities been consulted for the study?

The Ecopetrol engineer responded that the aim of the audencia was to discuss the technical aspects of moving the pipeline and replacing smaller pipes with larger ones in order to meet the increasing fuel needs of the region. He further noted that Ecopetrol was working with the Ministry of the Environment to ensure the ecological feasibility of the project, which he concluded would benefit all residents of the Pacific, including the black communities of Dagua.

Bismarck Chaverra, a Chocoan law student, rose next, carrying copies of the texts of the 1991 Constitution and Law 70. Referring to specific sections of each, he said that, as Colombian citizens with special rights, it was black communities, not just the Ministry of the Environment, who needed to be consulted and included in discussions about the proposed project. Before returning to his seat, he read part of Chapter II of Law 70, which decrees that the "participation of the communities and their organizations [be sought], without detriment to their autonomy, in all the decisions that affect them." Rosero followed Chaverra, arguing that infrastructure projects like the Ecopetrol pipeline were environmentally unsound and would thus destroy the resource-based livelihoods of black communities and violate a different section of Chapter II of Law 70—one that decrees "respect for the integrity and dignity of the cultural life of the Black Communities."

The Ministry of the Environment official reiterated the Ecopetrol engineer's claim that the state was committed to promoting sustainable economic development and that among the benefits of the proposed pipeline would be greater socioeconomic development for all Valle Caucanos. He noted that the forum, held at a public university, was open to everyone. Rosero and Aguilar countered that affected black communities were unlikely to participate in a forum held in Cali, not only because they lived several hours away, but also because they did not expect that views voiced in "everyday" terms or ideas based on their traditional, local knowledge of their world would be heard or heeded by technocrats and bureaucrats.

The exchange became heated and ended without resolution. On the following day, a small article on the pipeline ran in the Cali edition of the national newspaper *El País*, with no mention of the UniValle confrontation.

According to Donna Van Cott (2000: 5–6), the constitutional transformation in Colombia represents a commitment to "a set of democratic values with ethnic content influenced by indigenous cosmovision in which politics is embedded organically in a larger ethnic and cultural universe than the Liberal, Western, constitutional tradition of Latin America." Such claims and the passing of Law 70 notwithstanding, the recognition of black rights made little immediate difference in state policies and practices. The restructuring of the Colombian bureaucracy and the decentralization of administrative functions under the new constitution—and the paucity of knowledge about black culture, demography, and land-use practices—were partially responsible for such stasis. Perhaps more significant was the widespread ambivalence toward black ethnic and land rights. For a large number of state and development entities, like the ones at the Ecopetrol *audencia*, the path to a better future for the region and its inhabitants lay in political and economic modernization through plans such as Plan Pacífico.

While some black groups strove to take advantage of new political and economic openings in the country, PCN leaders saw the changes at the turn of the century as an opportunity to organize a broad-based "autonomous" black movement. In the post-Law 70 period, they attempted to rally diverse black groups around their identity as "Afro-Colombians" and to formulate cultural alternatives to mainstream economic models based on the practices of rural black communities of the Pacific. Although they failed in their efforts to gain sovereign control over the Chocó and have it declared a "collective ethnic territory," the notion of the Pacific region as a symbolic homeland for all black communities remained at the fulcrum of the PCN's ethno-cultural strategy. In an article co-authored with the anthropologist Arturo Escobar, the PCN's Libia Grueso and Carlos Rosero comment on black and indigenous movements in the region:

> Based on cultural differences and the rights to identity and territory, these social movements challenge the Euro-Colombian modernity that has become dominant in the region of the country. Black and indigenous cultural politics in this way, challenge the conventional political culture harbored in the practices of traditional political parties and the state, unsettle the dominant project of national identity construction, and defy the predominant orientation of development. (Grueso et al. 1998: 197)

The PCN and its ethno-cultural strategy bear close resemblance to new social movements, a term coined to describe the loosely organized coalitions of factory workers, peasants, women, urban squatters, and ethnic groups who arose in protests against the state and forces of late capitalism of the 1980s and beyond (Eckstein 1989; Escobar and Alvarez 1992; Jelin 1990; Redclift 1987; Rowe and Schelling 1993; Slater 1985). Marked by diversity of interests, identities, and organizing strategies, new social movements are anti-statist in that they neither seek inclusion in existing political structures nor participation in leftist revolutionary struggles to overthrow the state. Rather, these heterogeneous new social movements are seen as drawing on idioms of traditional or popular culture to resist the homogenizing forces of modernization and economic globalization and to imagine alternatives based on local knowledge. Although somewhat related to Western "identity politics," Third World new social movements are struggles as much for resources and democratic political participation as for the right to assert cultural difference (Cohen 1985). Echoing post-developmentalist positions, Arturo Escobar (1995: 217) characterizes such movements as grassroots attempts to "unmake development" and assert more creative, autonomous alternatives to it.

Inspired by the ideas of sustainable development and new social movements, I went to the Pacific lowlands of Colombia to study Afro-Colombian grassroots alternatives to economic development. My initial aim—to identify and formulate "generalizable" models of resource use in black communities' collective lands—was overdetermined by my disciplinary training in ecology and comparative politics. With my then recent engagement with "applied research" and nascent sense of politics, I hoped that such models would help promote sustainable development, conserve biodiversity, and assist Afro-Colombian groups to assert their ethnic and territorial claims.

But from the onset I ran into methodological and theoretical dilemmas. In addition to political ideology, black communities were marked by differences of region, local histories, class, and occupation, among other factors. Resource use and cultural practices varied widely and led to diverging opinions regarding which communities were most "representative" of Afro-Colombian identity and tradition. For example, many considered Chocoans to be the most representative of black communities because the Chocó is the only majoritarian black state in Colombia. Others asserted that the riverine communities along three southern Pacific coastal states epitomized traditional black practices, for they continued to log, mine, fish, farm, and engage in other subsistence activities using techniques established during slavery. There were also large numbers of Afro-Colombians in regional and urban centers in the Pacific, and in Andean cities, who had little connection to agrarian or rural livelihoods.

During my fieldwork, I observed activists, scholars, state and NGO officials struggling to come to terms with this heterogeneity and engaging in acrimonious debates over how to "operationalize" such terms in Law 70 as "black communities," "traditional practices of production," "collective territory," "mechanisms to protect [black] cultural identity," "participation without detriment to autonomy," and others.[11] Debates also raged about whether "blackness" or "Afro-Colombianness" was a racial or ethnic identity. The term "Afro-Colombian" came into circulation in the 1990s to stress the African descent of black communities and their connections to the Colombian nation-state. Despite disagreements over nomenclature and definitions, during my fieldwork the terms "black communities" and "Afro-Colombians" were most often heard and used interchangeably, as I do in this text.

Watching the dynamics of post-Law 70 struggles, it became evident that they could not be understood through "generalizable" models, waving away as anomalous the differences and inconsistencies within black communities and movements. In a first approximation (Asher 1998),[12] I saw black movements' attempts to construct a collective Afro-Colombian identity as a form of anti-developmentalist "cultural politics," wherein "culture is political because meanings are constitutive of processes that, implicitly or explicitly, seek to redefine social power. That is, when movements deploy alternative conceptions of woman, nature, race, economy, democracy, or citizenship that unsettle dominant cultural meanings, they enact a cultural politics" (Alvarez et al. 1998: 7). Escobar (1997, 1998, 1999) highlights how this cultural politics, combined with an appeal to biodiversity conservation and sustainable resource management, is a powerful discursive and organizational force through which black movements contest and engage the latest technocratic development interventions. Such cultural-political social movements, he suggests, offer real possibilities of an alternative modernity or development. In the case of the Colombian Pacific, he argues, struggles over identity and territory are also struggles over meanings and modes of production of nature/biodiversity (Escobar 1997: 221). Drawing on poststructuralist and feminist studies of science and technology, as well as political ecology frameworks, he speaks of "three different regimes of the production of nature—organic, capitalist, and technonature"—and hypothesizes about the utopian possibility of "hybrid natures," noting that "hybrid natures would take a special form in tropical rainforest areas, where popular groups and social movements would seek to defend through novel practices of organic nature against the ravages of capitalist nature, with technonature—biotechnology-based conservation and use of resources—as a possible ally" (Escobar 1997: 221–22). Escobar acknowledge that such a strategy faces immense obstacles.[13] Referring to black struggles in the Colombian Pacific, he elaborates: "The discourse of biodiversity and the dynamics of capital in its ecological phase open up spaces that activists try to seize upon as points of struggle. This dialectic posits a number of paradoxes for the movement, including the contradictory aspects of defending local nature and culture by relying on languages that do not reflect the local experience of nature and culture" (Escobar 1997: 220). Peter Wade (1995) also reflects on the dialectics and political economic context of the cultural politics of blackness. But unlike Escobar, he views the state as an important interlocutor and controller of black mobilization (Wade 1995: 350). He also differs from Escobar in that he sees the new political conjunctures as playing a constitutive role in shaping new black identities (Wade 1995: 353).

Escobar's reflections on the links between nature, culture, and the dynamics of capitalism, and Wade's focus on the structural dialectics of cultural politics, redirected my own observations of post-Law 70 black struggles. I noticed the double-edged effects that cultural legislation in Colombia has on its ethnic population (Correa 1993; Escobar 1995; Gow and Rappaport 2002; Gros 1991; Jackson 2002; Rappaport and Dover 1996; Wade 1995). On the one hand, the state grants black communities ethnic and territorial rights premised on their ability to prove that they are "ethnically different" from the rest of the population. On the other, the state attempts to incorporate them as citizens and members of civil society in a nation-state that had recently proclaimed itself multicultural. But like the parameters of black ethnicity, the meanings of multiculturalism and the boundaries of "nation," "state," and "civil society" are unclear and contested.

In contrast to Escobar, however, I found that contradictions lay not merely in expressing local realities through non-local language, but also in the complex and contingent ways in which the processes of black cultural organizing, state policies, and global development interventions in the Pacific were mutually constituting one another. Following Wade's reflections, these multiple contradictions also suggested the need to direct analytical attention to how political economic conjunctures, what black activists often invoked as the *coyuntura* of the Pacific, unfold in relation to Afro-Colombian cultural politics.

As my conceptual radar shifted, I noticed that, rather than autonomous expressions of resistance against the state (as PCN activists claimed or wished), black struggles, including understandings of local realities and culture in the 1990s, were at least partially shaped by and through the very discourses of political and economic modernity they opposed. For one, it appeared that both black movements *and* state interventions were using similar discourses and languages to construct their understandings of culture, nature, and development.

Despite their anti-institutional stance, PCN leaders used prevailing discourses of participation and new constitutional guarantees of citizenship rights to demand that the state recognize the tripartite Afro-Colombian claim of "identity, autonomy, territory." PCN activists did not seek formal political posts, but they did work in various capacities with state institutions involved in addressing Law 70 and with other development entities. Indeed, without such engagement the PCN would have risked de facto exclusion from important ongoing discussions of development and conservation.

PCN members held that their ethno-cultural proposal emerged from the perspective of grassroots communities. Yet concepts such as "local needs," "community participation," "rights of cultural groups," and "biodiversity conservation" were as central to the PCN as they were to the development and conservation rhetoric of private NGOs. Such notions were also mobilized by black groups other than the PCN. Black women's networks drew on the rhetoric of sustainable development and conservation initiatives and on discourses of "ethnic identity," "productive activities and relations with nature," and "everyday needs" in their efforts to organize as women and as blacks.

The shaping of black movements and identities by political-economic forces was hardly a one-way street. True, issues of environmental sustainability and the ethnic promise that Van Cott (2000) envisioned in the constitution were virtually absent from the initial drafts of Plan Pacífico. But state and economic planners faced immediate pressure to address social, cultural, and environmental demands. Such pressure came not only from ethnic movements and the language of the constitution and Law 70 but also from NGOs and state entities, some of whom were staffed by women and men genuinely committed to environmental and social justice. The imperative to make good on black rights by including the collective titling of ethnic lands as a component of the region's economic agenda also came from a seemingly unusual quarter, the World Bank. This advance was one of the contradictory elements of the globalizing phase of Third World development, what Jean Comaroff and John L. Comaroff (2001) call the "Second Coming of Millennial Capital." Advocates of neoliberal globalization (like developmentalism's predecessors) deem that market-led growth and "free trade" will lead to economic prosperity, social welfare, and environmental sustainability—that is, to efficient and sustainable development. Before retreating from the economy, however, governments in the Third World were required to undertake measures to liberalize trade from state control and regulation. Indeed, multilateral funding and development aid was contingent on such "structural adjustment" reforms, ushering in what Colombians call the *apertura económica* (economic opening). Plan Pacífico was a key component of Colombia's apertura.

In 1992, the World Bank approved a loan to fund a resource management project under Plan Pacífico with the condition that the titling of ethnic lands and preservation of knowledge about traditional resource use be included as components. Subsequently, a small percentage of the project's funding, and a great deal of the bank's attention, were dedicated to titling ethnic, including black, lands. As I discuss in detail in chapter 2, such an expansion of the project was in part a result of the intervention by the PCN, other ethnic groups, and NGOs during project negotiations. Similar interventions during discussions to implement the first phase of Proyecto BioPacífico and other conservation activities also led to the inclusion of ethnic concerns in environmental projects. But a celebration of these advances as the development industry's sensitivity to cultural and environmental issues, or even as an unconditional victory for ethnic groups, would be naïve. The collective titling of ethnic land was in keeping with neoliberal aims of clarifying property rights to reduce conflicts over resources and facilitate private-sector growth. Involving black and indigenous communities in resource management and biodiversity conservation is another attempt to address concerns over "local participation" and "environmental sustainability" while helping to integrate these heretofore marginalized populations into the market.

To see such developments merely as capital's co-opting of black struggles would be equally uncritical, however. In the past decade, several Afro-Latin American groups have taken advantage of the broad interest in "Afro-descendent" communities to seek support from multilateral development banks for their land rights and other democratic guarantees promised under the new constitutional reforms (Inter-American Foundation 2001; Offen 2003; Thorne 2001). In other words, the "*experiential* contradictions of neoliberal capitalism" and the "uneasy fusion of enfranchisement and exclusion" that Comaroff and Comaroff (2001: 8) observe elsewhere are evident among Afro-Colombians and other Afro-Latino groups. One need not be an apologist for globalization to note that, far from being uniformly homogenizing, its effects are double-edged, contradictory, and circumscribed by local history and culture.

Near the end of the twentieth century, ethnic rights and traditional knowledge (along with sustainable resource use and biodiversity conservation) became key idioms through which modernity and development began to be asserted and understood in the Colombian Pacific and many other places in the Third World. Such expressions of development had another, seemingly paradoxical effect: at a moment when neoliberal globalization measures and the emphasis on private, market-led economic growth called for a decrease in the role of the state, they played a major role in helping the Colombian state gain legitimacy and constitute itself. During the mid-1990s, many state and subsidiary agencies appeared in a region where they had had little presence or influence to promote development, environmental conservation, and the recognition of ethnic and territorial rights. Such state legitimacy was partial and weak, at best, as the increasing presence of drug traffickers, paramilitaries, guerrillas, and other extralegal armed groups in the Pacific in the last decade bears tragic testimony. Nor could legal recognition guarantee the realization of rights. Even as the first collective land titles were handed out in 1997, black groups began to be displaced from the Pacific in vast numbers because of the dramatic escalation of violent conflict. These newly vacated lands were quickly taken over for illicit coca cultivation and commercial monocultures (such as African oil palms and shrimp ponds). The structural violence of such penetration endangers the already precarious local access to land and livelihoods. Indiscriminate aerial fumigation as part of the "war on drugs" also destroys subsistence crops and natural vegetation. As the "war on drugs" morphed into the "war on terror," it paved the way for increased state-sanctioned violence in the name of "peace and security."

The reality of the Pacific and the black movements is very different today from that of the mid-1990s. The bleak state of affairs in the Pacific can be seen as evidence that the project of the black communities against state modernization and globalization failed. But such a conclusion only emerges from and reinforces a false binary—that Afro-Colombians in particular, and local groups in the Third World in general, are neglected or overexploited victims of development or heroes of local, cultural alternatives. It would also shortchange the gains for Afro-Colombian groups: black representation in formal politics and various state institutions, collective land titles given to about 122 communities by 2003, an increase in grassroots organizing, and greater awareness and support for the struggles of black communities in transnational movements.

How can one explain the dynamics in the Pacific adequately to account for the ways in which local, national, and global processes are intertwined? How can one account for the interplay of power relationships at work at and across these levels? How can one be in critical solidarity with black struggles while eschewing apolitical explanations of globalization and romantic understanding of social movements? To address these questions, I draw on work that considers the processes and crises of capitalist development and associated struggles for social change in historical and analytical perspective.[14]

Early critiques of development focused on structures of the inequalities and material consequences of development. Works influenced by postmodern currents show how these material effects function through specific discursive productions and representations of the Third World and its peoples as "underdeveloped" objects and subjects in need of intervention (Escobar 1995; Ferguson 1994; Mitchell 1991).[15]

Drawing on Michel Foucault's (e.g. 1980, 1983, 1991) work on power/knowledge, both James Ferguson and Escobar trace how development operates as a "regime of truth." Ferguson argues that, although rural development in Lesotho "failed" on its own terms, it successfully produced a discourse—a set of institutions and practices—about development, which had important, unintended effects. According to Ferguson (1994: xv), the development "apparatus . . . is as an 'anti-politics machine,' depoliticizing everything it touches, everywhere whisking political realities out of sight, all the while performing, almost unnoticed, its own preeminently political operation of expanding bureaucratic state power." Escobar (1995: 9) also sees the development discourse as having "profound political, economic, and cultural effects that have yet to be explored." He postulates that development was invented in the decade after World War Two but is an extension of "the colonial move."[16] That is, it is governed by the same core impulse as colonialism in which the West creates "an extremely efficient apparatus for producing knowledge about, and the exercise of power over, the Third World" (Escobar 1995: 9).

Escobar recognizes that development takes varied forms across time and geopolitical contexts and that a multiplicity of mechanisms and subjectivities are at play in inscribing and maintaining its power. He applauds theories of imperialism, unequal exchange, world systems, and peripheral capitalism for showing how a global system of economic and cultural production appropriates local bodies of knowledge and resources. However, he finds these political economic inquiries inadequate for not bringing "to the fore the mediations effected by local cultures on translocal forms of capital" or investigating "how external forces—capital and modernity—are processed, expressed, and refashioned by local communities" (Escobar 1995: 98).

Escobar follows his deconstructive move with a reconstructive one by drawing on Latin American social theories of "hybrid" cultures and modernity. He sees new social movements from the Third World bringing about "the end of development as a regime of representation" by displacing "normal strategies of modernity" and producing "different subjectivities" and hybrid alternatives to "the Western economy as a system of production, power and signification" (Escobar 1995: 216–17).

While I agree with Escobar about unmaking the power of development and his call for a more culturally grounded political economy, I differ with his analysis and post-developmentalist solution. For one, viewing development as an extension of Western colonialism is at odds with an important insight of postcolonial and new marxist scholarship—namely, that the "West" and the "rest" emerge relationally and constitute each other in material and discursive terms. Although post-World War Two developmentalism and its retinue of Bretton Woods institutions established "a new mode of global governmentality" (Gupta 1998), the development project is filled with ambiguities, not least because of its linkages with other discourses such as nationalism (Gidwani 2002). The search for alternatives to capitalist development and modernity, then, needs to begin by examining how development and anti-developmentalism are implicated within one another (Watts 1995).

Escobar and other anthropologists are surely right when they insist that local communities in the Third World not only contest and resist global forces of development but also refashion them. But these communities and the social movements that emerge from them are not simple manifestations of radical, non-Western culture, as post-developmentalist discourses of difference imply.[17] Rather, the interconnections between structural forces and divergent discourses (including those of nation-state formation, or, more recently, globalization and environmentalism) are important elements in understanding how local identities and interests are shaped and articulated (Gupta 1998). That is, globalization's "cultural and material forces do not have simple homogenizing effects. They are in some measure, refracted, redeployed, domesticated, or resisted wherever they come to rest" (Comaroff and Comaroff 2001: 14). Such refractions and refashioning form and transform both "local" and "global" forces, albeit in uneven ways. While I remain deeply critical of the promises of state modernization and globalization, I argue that it is important not to underestimate the degree to which development and resistance are related dialectically.

Even the new social movements' notion of "hybrid" models of resistance does not sufficiently call into question the teleological direction of development practices and static or "given" representations of traditional and popular culture. It implies that local communities incorporate or adapt "modern" technologies while leaving the apparatuses of modernization and tradition intact. Such a position tells us little about how development practices may be transformed through applications in different locations and historical moments—or, more centrally to the Pacific case, how local strategies of resistance such as black cultural politics are shaped through their active engagement with, not just against, the development and democratic practices of the Colombian state. By this I do not mean to suggest that black cultural politics are "structural effects," examples of "strategic essentialism," or overdetermined outcomes of external forces. Rather, I take seriously Akhil Gupta's suggestion that "it is important to maintain the tension between the universalizing and globalizing power of development and its disputed or contentious redeployment in particular cultural and historical locations" (Gupta 1998: 16). Black women's networks are a case in point: members drew on discourses of ethnic and gender identity, sustainable development, and biodiversity conservation to advance their own agendas while at the same time being constituted as subjects of development.

Feminist insights are crucial if we are to understand the dynamic nature of domination and resistance and the uneven and multiple power relations within which women act.[18] Stressing the heterogeneity of women's and feminist movements, recent work in Latin America traces how race and ethnicity intersect with gender, class, and other factors to shape women's subjectivities, needs, and activism.[19] Feminist and postcolonial theorizing also serves as a reminder to reflect critically on our desires and methods to better the lives of Third World women, and Third World peoples in general, and to engage in more inclusive and strategic politics to bring about social change.

Black activists, especially PCN leaders, were keenly aware of the paradoxes of development's power and exploited them through the possibilities offered by local, cultural, and, indeed, national conditions. They were equally conscious of the difficulties of organizing for change across their differences and the possibility that the political and economic alternatives they imagined for the Pacific would not come to fruition in the near future. Yet under the shadow of what was coming—perhaps in the hopes of pre-empting it—they seized the moment that the new constitution and Law 70 provided. In this spirit, they embarked on a process of constructing a movement of "black communities" and of imagining an Afro-Colombian cultural utopia based on the right to be black and different and to have symbolic and material control over land and resources. This process was neither autonomous nor removed from the project of Pacific modernity and development. Rather, it was located firmly within the contradictions and aporias of capitalist development, which must be understood as part of the long-term task of unmaking their destructive effects.[20]

Plan of Book and Methodological Remarks

This book examines the dynamics of black movements and their engagement with broader political economic processes in the mid-1990s in some detail, before returning to their implication for understanding development, modernity, and social movements. Chapter 1 describes the emergence and development of the movement to include black rights in the 1991 Constitution. In that section of the book, I also discuss how ideas of race, culture, and citizenship, and especially the changes in the regional political economy, influenced the nature and extent of black legal rights granted in Law 70. The chapter ends by remarking on the splits in post-Law 70 black movements and how the changing dynamics of the Pacific are key to subsequent black organizing.

Chapter 2 focuses on the state's vision of development for the Pacific region and how ethnic concerns become part of the state's economic agenda. The chapter describes three state-administered projects—Plan Pacífico, an economic development plan; Proyecto BioPacífico, a biodiversity conservation project: and *Ordenamiento Territorial*, a territorial zoning policy. It discusses how black interests are accommodated in the development agenda and how the notions of economic development, ecological sustainability, and ethnic rights help establish and expand the state's power in political and cultural terms. In chapter 3, I take a detailed look at the PCN's attempts to build a unified black movement based on the call for "identity, territory, autonomy" and the difficulties of taking into account the heterogeneity (of history, region, class, gender, variable cultural practices, and political ideology) of black communities in constructing a common political agenda. It continues with a discussion of the PCN's failed efforts to develop a culturally appropriate Development Plan for Black Communities (Plan de Desarrollo de Comunidades Negras; PDCN) and the issues and challenges of grassroots-level organizing. The chapter ends by discussing how the heterogeneity of black identities and interests cannot be understood in binary terms such as "equality versus difference" and cautions as much against romanticizing resistance as against reducing social struggles to structural effects.

Chapter 4 explores how Afro-Colombian women's organizations and networks engage regional development initiatives and ethnic struggles. It discusses how the gender policies and politics of sustainability of development institutions shape Afro-Colombian subjectivities and organizations and how black women, in turn, draw on the terms and resources of development initiatives and cultural discourses prevalent in the region to address their needs. By tracing the changing dynamics of Afro-Colombian women's activism, I show how it was shaped through black women's active engagement with, and against, the development practices of the state and of black ethnic movements.

The concluding chapter outlines how Afro-Colombian (now referred to as Afro-descendant) movements and development policies changed as the latent national crisis and low-intensity conflict escalated into a full-fledged civil war. Local communities in the Pacific were increasingly affected by the armed offensives between extralegal, paralegal, and state forces that spread rapidly into the Pacific region. With respect to the Afro-Colombian communities, there is a burgeoning in the number of black groups, further fragmentation and polarization among them, more linkages to mainstream politics, massive out-migration from the Pacific, and stronger alliances with transnational social movements. I conclude with reflections on the conjunctural nature of the relations between black cultural politics and political economic processes in particular and social movements and development processes in general.

This book is grounded in fifteen months of ethnographic fieldwork conducted in Colombia from 1993 to 1995 and one five-week trip in 1999. Law 70 had to be implemented by 1995 (within two years of its ratification in 1993). Between 1993 and 1995, numerous fora, symposia, seminars, rallies, conferences, and public and private meetings were held to discuss issues of black rights and the Pacific. I was a participant observer at many of these events in Bogotá, Cali, and various locations in the Pacific. There I had a chance to witness cultural politics in action as different factions articulated their positions and engaged in heated debates. Participating in such meetings also prepared me for interviews and conversations with members of black movements, grassroots organizations, women's cooperatives, state officials, and NGOs involved in implementing ethnic, development, and environmental plans. At these meetings I followed up on issues raised during formal events or asked for elaboration. At a meeting convened by the Ministry of the Environment similar to the Ecopetrol forum, for instance, I saw how both state officials and black activists mobilized discourses of biodiversity conservation to assert their agendas. During the exhaustive discussion, the state and NGO officials and black activists overlapped, diverged, and angled for compromises. Thus, participant observation at public events and movement meetings, conversations, and interviews were key ethnographic approaches of my research. I conducted structured and unstructured interviews and had extensive conversations with members of black movements and women's cooperatives in the Pacific region. I also spoke with members of various state entities and NGOs involved in the implementation of Law 70.

Because much of what was communicated to me in interviews appeared later in printed sources or was reiterated at public events, I have for the most part chosen to use the real names of my informants. Wherever possible, I have cited these sources. Where their comments might have negative consequences for my informants, I have not indicated their identity.

From the onset, my subjects' identities and ideological views (as well as my own) posed crucial methodological and epistemological dilemmas. My training as a biologist in post-independence India and as a political scientist at an American university stressed the importance of "objective distance" from research subjects. My identity, my political beliefs and prior knowledge, the context within which the research was done—none of these were supposed to have any bearing on my research. However, any attempt to claim that my research was neutral and objective would be futile. During the particular constellation of cultural politics unfolding in Colombia in 1995, my identity was always already present, a shifting but undeniable signifier in every context.

As an Indian, I was invariably asked about my politics and position regarding Gandhian nonviolence, anticolonialism, and India's nonaligned status during the Cold War. A fluent Spanish speaker who lived in the southern United States and had done extensive fieldwork in Latin America, I was often assumed to have a "natural" solidarity with black struggles—which I did. Because of my identity and politics, I was considered something of an insider and was variously invited to observe, expected to attend, or graciously tolerated at the many meetings held by and for black communities. However, because I was not quite structurally black, my presence and participation at these events were viewed with a certain ambiguity. At one PCN retreat to which I was specifically invited, I was asked to teach yoga to help activists relax after a long and frustrating workshop on developing political strategies—a workshop I had not been allowed to attend.

Because of my insider-outsider position, most activists from regional movements, including those whose positions were at odds with each other, were willing to talk to me and respond to my questions about their views on Afro-Colombian struggles. My approach became influenced by Gayatri Spivak's notion of moving beyond neutral dialogues "to render visible the historical and institutional structures of the representative space" from which one is called on to speak (Spivak 1990: vii).

The multiplicity of gender, region, and ideological factors that marked black ethno-cultural politics enabled and shaped certain conversations while blocking and frustrating others. For example, black women leaders and black women's groups often struggled with their crucial yet unacknowledged role in black political struggles. Nor did black women always agree with their compañeros about who was or was not an "insider." Among some Afro-Colombian women's groups I was treated as a "double insider"—as a woman and as non-white. Many black women were especially keen to discuss their experiences with ethnic and gender struggles and to ask me to share my (then rudimentary) knowledge of Indian women's struggles. While my ambiguous insider-outsider status did not give me "objective data" on the position of black activists, it allowed me to gain insights into the multiplicity and contradictions of their ethno-cultural politics.

I also made short trips (usually a week) to riverine and coastal communities in the four Pacific states and one extended field stay (two months) in the village of Calle Larga in Río Anchicayá, Valle del Cauca. These field visits gave me a glimpse of the diversity and complexity of black self-identity, cultural and resource-use practices, and the norms governing collective land ownership. I also observed how the rapidly changing political economy was affecting grassroots communities in the region and gained a firsthand understanding of the communities' perceptions and engagements with development and conservation projects and the PCN's ethno-cultural organizational strategy.

During the 1990s in Colombia, there was a burst of "gray literature" and publications—academic, institutional, and popular—on various aspects of black rights and on economic and environmental concerns in the Pacific region. In addition to my direct observations, I draw extensively on many of these sources. My work especially relies on the plethora of unpublished material—reports, minutes of meetings, statements of objectives and principles, informal video recordings of their meetings—generated by black organizations and women's groups. I collected much of this material during short trips to Colombia in 1998, 1999, 2004, 2005, and 2007. During these trips I also had crucial conversations with PCN activists, many of whose key members have now been displaced from the Pacific to Bogotá because of the violence and threats to their lives.

✳ 1

AFRO-COLOMBIAN ETHNICITY

From Invisibility to the Limelight

After Brazil, Colombia has the largest black population in Latin America. But it was only recently that both countries conferred specific rights to their black populations—or, indeed, recognized blacks as a distinct group within the nation. Why this is so is linked to a number of factors—histories of nationalism, the perception of blacks and blackness within prevailing ideologies of "race" and culture, and the structures and dynamics of political economy—and how they function in each context. There is an extensive and growing scholarship on the often "hidden" histories and complex interactions of "race," culture, and nationalism.[1] An in-depth engagement with this literature is beyond the scope of this study, but a brief overview follows to contextualize current black social movements in Colombia.

According to Norman Whitten and Arlene Torres (1998), three nationalist ideologies—racial mixture, Indianism, and blackness—functioned in the Spanish-, Portuguese-, and French-speaking American republics in the 1990s. In much of Latin America, the ideology of race mixture (*mestizaje*) between Europeans, indians, and blacks has prevailed since the post-independence period. Indianism (*indigenismo*) functions as a two-pronged component of mestizaje. It is seen as providing an authentic basis for a distinct Latin American nationalism but also as responsible for the "backward" elements in an underdeveloped nation. The third nationalist ideology—blackness, or négritude—takes as positive the power attributed to people identified as "black" and was only adopted by Haiti.[2] Whitten and Torres note that nationalist ideologies not only develop and depend on symbols of unity (often understood in terms of "race," blood, or biology) but also of difference (understood in "cultural" terms).

Colombia, like many countries of Latin America, portrays itself as a nation of mestizos. Peter Wade (1993a) notes that historical mestizaje was a complex process and a contradictory ideology. While it celebrated the diverse elements of racial mixture, European or white components were nonetheless coded as more civilized and modern and, hence, more valuable. The process of miscegenation was intended to produce a homogeneous people speaking one language and believing in a single god (de Friedemann and Arocha 1995; Whitten and Torres 1998) and implied the whitening (*blanqueamiento*) of its darker populace and its eventual integration and assimilation into mainstream Colombian society. Indeed, mestizaje and blanqueamiento, common throughout Latin America, were supposed to help neutralize forms of diversity considered subversive, challenges to the official nation.

Anthropologists claim that these nationalist ideologies and the underlying structures of the colonial and postcolonial political economy are largely responsible for the "invisibility" of racial, cultural, and ethnic differences in Colombia and for the socioeconomic marginalization of minority groups (de Friedemann and Arocha 1995; Wade 1993a, 1995; Whitten and Torres 1992). But the dynamics of color, culture, and class played out differently for indians and blacks. These differences have had major implications for land rights.

Colombia has a long history of viewing indigenous communities as culturally distinct and recognizing—or, at least, legally articulating— their special rights (Gros 1991). For example, Law 89 of 1890 granted Colombian indians collective title to their lands and recognized the traditional authority of indigenous councils (*cabildos*) to govern and manage affairs within such territories. This law and other recognitions of difference were an extension of indigenist colonial policies to keep indians separate from whites or creolos. Concentrating indians in reserves (*resguardos*) also ensured the availability of native peoples as sources of labor and tribute. Resguardos and cabildos are classic examples of indigenism. They were seen as protecting the purity of indian culture (and providing labor reserves) but also ensured that indigenous "cosmovisions" did not contaminate the liberal individualist political philosophy of the nation-state.

Indians found opportunities to deploy this ideology for ends of their own. When agrarian reforms in the 1950s and 1960s threatened to dissolve resguardos, indigenous groups (and their anthropologist and lawyer advocates) used indigenist arguments to resist forced assimilation and appropriation of their land. They drew on the notion of tradition and custom and under Agrarian Reform Law 135 of 1961 were granted autonomous collective control over the lands that they had traditionally occupied.[3] Since the 1960s, indigenous organizations in Colombia have invoked various international and national accords on the rights of cultural minorities to demand legal and administrative rights to self-government.

The case was different for blacks. While the exact numbers are subject to debate, it is estimated that 10 percent to 30 percent of the country is of African descent.[4] The vast majority of Afro-Colombians live in the Pacific littoral, where they constitute 90 percent of the population. A significant percentage of this population is concentrated in urban centers (Urrea Giraldo et al. 2001) and increasingly being displaced to the Andean part of the country. The postcolonial constitution of 1886 defines blacks as mestizos, subject citizens with ostensibly the same political and economic rights as all other Colombians (except indians).

After manumission in 1851, blacks scattered along the length and breadth of the Pacific region, often joining existing *palenques*—settlements established by escaped or freed slaves (*cimarrones*). As black communities expanded and spread in the Pacific (and some other areas), Afro-Colombians developed a distinct set of cultural-symbolic beliefs and material practices, combining elements from their African past with new features developed in their present circumstances. The anthropologists Nina de Friedemann and Jaime Arocha (1984, 1986, 1995) stress that Afro-Colombians chose to isolate themselves from mainstream society as much as an act of resistance and independence as to escape racial discrimination and persecution.

Peter Wade (1991, 1993a) interprets the situation of black Colombians differently. He argues that the nationalist processes of mestizaje and blanqeamiento had a twofold effect. Some populations, especially along the Atlantic coast and in urban areas of Antioquia, assimilated culturally within dominant society, albeit subject to racism and discrimination. Others, such as those in the Pacific region, remained isolated from mainstream society more by imposition (as a structural effect of the political economy) than by choice. According to Wade, the fundamental contradictions within Colombia's national ideologies—foreseeing an ultimate homogeneity but insisting on hierarchical distinctions based on "race"—led to a complex coexistence of integration and discrimination for blacks.

Roque Roldán (1993) asserts that in jurisdictional terms, blacks living in rural areas were treated as any other Colombian *campesinos* (peasants): all faced the same problems of land tenure and socioeconomic development. In his study of black peasants of the Cauca Valley in the late nineteenth century, Michael Taussig (1980: 58) describes his subjects as "outlaws—free peasants and foresters who lived by their wits and their weapons rather than by legal guarantees to land and citizenship." In the twentieth century, the hacienda economy broke down, and agricultural production in the region became increasingly organized according to capitalist logic. As Taussig's ethnography shows, even as these black outlaw peasants became incorporated into the wage economy as laborers, they drew on "black cosmology" to interpret and denounce the capitalist logic of agribusiness. They also continued to engage in non-market subsistence practices and retained social relations among black kin in the Pacific coastal region. Black communities also coexisted peacefully with their indigenous neighbors and often established kinship ties with them.

Whatever the understanding of the black situation, black communities had no special land rights. Nor were they legally considered a culturally distinct group until the inclusion of AT 55 in the 1991 Constitution.

During my fieldwork, I heard diverging and fragmented versions of how this legal victory was attained. For many ordinary Colombians and state officials, the recognition of black ethnic rights was an extension of the change in Colombia's nationalist ideology. Some referred to Article 7 of the 1991 Constitution, which declares Colombia to be a multiethnic and pluricultural nation. But many who espoused this view conceded that the state was reluctant to grant special ethnic or cultural status to black communities.

During our conversation in April 1995, Angela Andrade, subdirector of the Agustín Codazzi Geographic Institute (Instituto Geográfico Agustín Codazzi; IGAC), told me another story about the origin of the collective-title clause in AT 55. In 1991, she flew over the Pacific region in a helicopter with the sociologist Orlando Fals Borda. During the flight, Fals Borda was struck by something he had noticed on an earlier flight: almost the entire area was unfenced. After discussing this issue with local communities in the region, he concluded that they managed resources in a collective fashion and that people did not hold individual titles to their lands. Fals Borda, one of the framers of AT 55, consequently saw fit to enshrine collective land rights for black communities in the draft legislation. Andrade notes that Fals Borda later acknowledged that the land situation in the Pacific was more complicated than he thought and that by stipulating collective land rights as the basis of AT 55, he had inadvertently opened a can of political and administrative worms.

Fals Borda's mea culpa overstates his influence: the can was pried open by many. Legal recognition of black rights was a controversial affair riddled with conflicts and opposed by powerful actors in Colombian society and the state. Black activists, not surprisingly, emphasize the role of their own organizing and mobilization efforts in bringing black demands to national attention. Before assessing such a claim, an overview of pre-1993 black organizing—and the cultural, political economic, and ecological context in which it happened—is in order.

Land Struggles in the Northern Chocó:
The Seeds of AT 55?

In addition to substantial representation in the Colombian metropolises (Bogotá, Medellín, and Cali), there are six regions of important black presence in Colombia: the Atlantic/Caribbean coast; the Magdalena river valley; the Cauca river valley; the Patía river valley; the San Andrés and Providencia archipelagos (where English predominates); and the rural riparian zones of the Pacific littoral.

As noted previously, the Pacific littoral is the largest contiguous area of black presence in Colombia. As such, it has played a crucial role in shaping notions of Afro-Colombian identity. In the colonial era, Spanish invaders decimated or dislodged the vast majority of the region's indigenous inhabitants, replacing them with African slaves brought to work in the gold placer mines. Through the collapse of Spanish rule in 1810 and the arrival of manumission in 1851, blacks and the surviving indigenous peoples formed the backbone of the region's political economy.

Escaped and freed blacks continued to arrive and settle in the Pacific lowlands throughout the nineteenth century.[5] Alongside Emberá and Waunana indians, these newcomers farmed, fished, hunted, mined, logged, and engaged in other subsistence activities. Over decades, culturally vibrant black settlements developed along the extensive river valleys and the littoral. Yet the Pacific remained subject to capitalist economic forces. Particularly transformative were the various boom-and-bust cycles of resource extraction. Both forest products (timber, rubber, tagua, etc.) and minerals (gold, platinum, silver) provided lucrative trade for outsiders (de Friedemann and Arocha 1986; West 1957; Whitten 1986).

Under Law 2 of 1959, vast areas of the country, including extensive swathes of the Chocó, were declared state forest reserves and *tierras baldías* (empty or uninhabited lands). While the Chocoan indians had a semblance of control over their communally owned resguardos, black inhabitants of these rural zones became de facto squatters, or *colonos*. The Pacific economic "frontier," at the geographic and economic periphery of the Andean centers of commerce, also became the focus of major development interventions in the 1950s. As I discuss in the next chapter, these interventions intensified as a result of Colombia's apertura in the 1980s. While the "invisible," "ignored," or "exploited" Afro-Colombian communities were not key targets of these interventions, their mere presence required treatment beyond the longstanding habit of neglect. Moreover, given their own ambitions, these communities became central actors in the unfolding drama in the region.

While there are significant black communities in all four Pacific departments, only Chocó has a black majority (in the other three, non-blacks in the interior outnumber coastal blacks). Since the middle of the twentieth century, a black elite based in Chocó's capital, Quibdó, has dominated the department's political administration, including the Chocó Regional Development Corporation (Corporación Autonóma Regional para el Desarrollo del Chocó; CODECHOCO). Rural communities and other poor groups remained on the margins of this patron–client system of politics and were exploited within the parameters of an extractive economy. In the 1980s, the precarious livelihood of black peasants came under increased pressure as CODECHOCO handed large concessions to private logging and mining firms. With the help of the Catholic church, black peasants in the Atrato region organized to contest these giveaways (Arocha 1994; Pardo 1998; Wade 1995).

It was during this period that the United Peasant Association of the Atrato River (Asociación Campesina Integral del Río Atrato; ACIA) emerged as one the largest associations of black peasants in the Chocó. In a March 1995 conversation with me, ACIA members recounted the history of their group's formation:

> The March to June harvest time was very wet, and our harvest was destroyed. In difficult times we used to get loans from the Caja Agraria [a state-run agrarian financial institution]. But they asked for 30 million pesos as a guarantee deposit for the loan we requested.
>
> We organized together with the missionaries and grassroots groups to fight the logging concessions given to Maderas de Darién, later Balsa II and

IV [all logging interests]. But we got discouraged with praying on an empty stomach and severed our church connections. We got our *Personería Jurídica* [legal status] and became an NGO with the aim of protecting our natural resources.

Esperanza Pacheco, the very nice lawyer who is the asesora [adviser or consultant] for OREWA [Organización Regional Emberá–Waunana del Chocó, the regional organization of the Emberá indians] advised us, and we realized that indigenous people and black communities have similar problems. We felt that this problem of land was not just a problem for us as peasants but also a problem of black communities.

In June 1987, ACIA organized the first forum in Buchadó, Chocó, to discuss land and forestry issues in the Atrato region (Agudelo 2001). Present at the forum were officials from several federal agencies, including the National Planning Department (Departamento Nacional de Planeación; DNP), Institute for the Development of Renewable Natural Resources (Instituto Nacional de los Recursos Naturales Renovables; INDERENA), and the Agrarian Reform Institute (Instituto Colombiano de la Reforma Agraria; INCORA), and the regional directors of CODECHOCO. The forum ended with an accord between CODECHOCO and ACIA that granted local communities collective usufruct rights over 600,000 hectares. Under Law 2 of 1959, this area was part of a state-owned forest reserve, and ACIA communities were required to employ sustainable forestry management practices when using natural resources. A few months later, ACIA and CODECHOCO signed another agreement that expanded these rights to 800,000 hectares (Agudelo 2001). Despite the accords, local communities continued to face problems with lack of funding and technical assistance, difficulties demarcating what area could be used for what purposes and COCECHOCO's refusal to abide by its own terms.

In response, ACIA broadened its base and refined its strategies. The group continued to draw on the experience of OREWA, which had been working to gain and expand control over its communal resguardos since the 1970s. ACIA also joined popular organizations such as Quibdo's Organization of People's Neighborhoods and Black Communities of the Chocó (Organización de Barrios Populares y Comunidades Negras de Chocó; OBAPO) to draw attention to the neglect of poor black communities and gain visibility for their social and economic struggles. In one instance, several grassroots organizations engaged in a peaceful takeover of the mayor's office in Quibdó.[6]

When CODECHOCO refused to back down, ACIA took its demands to Bogotá. Federal officials did meet with ACIA to discuss the problems of logging concessions on this occasion (Pardo 1998; Pardo and Álvarez 2001). A delegation of federal officials and representatives of CODECHOCO also met ACIA and its allies (including missionaries and representatives from NGOs, indigenous groups, and other grassroots communities) in August 1988 at a forum in Padua, Chocó. Representatives from DIAR, a Dutch-funded agro-forestry initiative in the region, were also present at the meeting.

In Padua, ACIA demanded that its claims over land that black communities had traditionally occupied and used sustainably be recognized and that the communities be given title to such lands. Carlos Agudelo (2001) argues that ACIA's earlier claims for property were classic peasant demands for land. However, probably inspired by its alliance with OREWA, ACIA henceforth couched its claims in ethnic and environmental terms—language that would become increasingly significant in the ensuing black struggles.

First, ACIA argued that, according to International Labor Organization accords on the rights of ethnic minorities, black communities had the right to use, manage, and collectively own the lands they occupied. Unlike indigenous communities, they could not make claims to originary rights over this land. However, as one ACIA member said to me at our 1995 meeting, black communities had rights to land *por transcendencia*—that is, as a consequence of having settled and lived in the region for almost five hundred years.

Second, they linked their land claims to the state's prevailing policies to promote "sustainable development" or "green" resource management in the Pacific region. ACIA members argued that the economic or subsistence livelihood strategies of black communities were environmentally friendly and that granting them land and resource rights (as opposed to concessions to private firms) would aid the state's mandate to promote sustainable development. DIAR officials attested that local communities engaged in "ecologically sustainable" forestry practices and supported ACIA's proposal for land ownership and communal resource management. ACIA members also argued that their economic practices were based on the specificities of black cultural and social life. Thus, they added a social component to sustainable use and management arguments that had heretofore focused on technical (ecological and later economic) aspects of resource use. I will return to the issues raised by all three claims presently.

ACIA's efforts bore mixed fruit. Its members were not granted land titles; nor were their ethnic claims recognized. However, federal officials reiterated that commercial logging would not be permitted in areas set aside in the Buchadó accord. Despite these guarantees, commercial logging operations in the region continued.[7] Critics argued that commercial logging in areas slotted for communal management occurred with CODECHOCO's blessing. CODECHOCO officials denied the charge but at the same timed defended commercial forestry as a legal and necessary part of broader economic development in the region. They further claimed that these operations complemented (rather than harmed) the economic activities of local communities.

In 1989, ACIA, the Peasant Association of the San Juan River (Asociación Campesina del Río San Juan; ACADESAN), and indigenous groups organized a bi-ethnic meeting in the region. According to Mauricio Pardo (1998), a proposal to establish an immense bi-ethnic territory in the lower San Juan region was aired at this meeting for the first time. While no such territory was established, the meeting provided ACIA with yet another opportunity to broaden its alliances. Such encounters, impelled by the deepening land struggles in the northern Chocó, often served to unite the efforts of previously disparate groups. Immediate aims may not have been met, but without such gatherings and the work of assertive organizers in the Río Atrato area, black rights might never have found mention in Colombia's new constitution. ACIA's demands and activism in particular were crucial precursors to the eventual debate over AT 55.

How AT 55 Became Part of the 1991 Constitution

While arising from specific socioeconomic conditions in the northern Chocó, the struggles of ACIA and other peasant groups were not aberrations. As mentioned in the introduction, from independence forward Colombian political history has been marked by a series of violent conflicts and a multitude of protests against the inability of the Colombian government to meet people's basic needs. Ruling elites (from both political parties) were not interested in sharing power with the poor and ignored their concerns (Palacios 2006). The rural poor were particularly marginalized, and poor rural dwellers became targets of political violence during the decade of La Violencia (1948–1958). The Pacific region remained on the margins of this particular wave of armed violence and the guerrilla activities that emerged subsequently (in the 1960s). Nonetheless, the poor in the Pacific suffered from state neglect. Neither the state's half-hearted land reforms policies nor its agricultural modernization measures (both undertaken in the middle of the twentieth century, in the 1960s) significantly benefited poor rural communities in the Pacific. Black and indigenous groups were among those who protested against this neglect and were active in varying degrees in the National Association of Peasant Producers (Asociación Nacional de Usuarios Compesinos; ANUC). For these groups, poverty combined with ethnic or racial difference to magnify their marginalization, albeit in different ways. Since the 1970s, various indigenous organizations and movements had arisen to assert cultural and territorial rights (Dover and Rappaport 1996; Findji 1992; Gros 1991; Jackson 2002). Black groups also organized around their particular concerns, as I discuss below.

The constitutional-reform process that began in the late 1980s was a response to both accelerating violence and increasing socioeconomic and political marginalization of large sectors of the populace. The reform process involved grappling with many issues—yet another peace accord that included the complex and incomplete process of demobilizing guerrilla and other armed groups, economic *apertura*, environmental conservation, the restructuring of the state, the reconstituting of civil society, and changing notions of citizenship and national culture. A National Constituent Assembly was elected to change the constitution and help usher in a more peaceful national era where the needs of all Colombians would be met. Numerous peasant and aborigine groups, including Afro-Colombians, also brought their concerns to this forum. They articulated their claims in terms of territory and administrative autonomy and linked them to concerns over environmental conservation and cultural rights (which were fast becoming part of development lexicons).

There is little evidence that ethnic or cultural rights appeared on the Constituent Assembly's list of concerns before black and indigenous activists pressed the organization for their inclusion—from the outside, during the Constituent Assembly elections, and later as a result of indigenous representation within it. Whether forced or reluctantly persuaded to consider ethnic rights, the initial response of the Constituent Assembly was to attempt to address them (and particularly black rights) as a subset of its modernizing, state-building project via an extension of citizenship rights (Agudelo 2001: 12). Indeed, the Colombian state revealed a marked preference for assimilating the rights of black communities within its neoliberal economic agenda, principally through the delineation of property rights. Rather than claim what most considered an unhealthy compromise, however, black groups from the rural Pacific steered a complex obstacle course to push for territorial and administrative autonomy along the lines of what was being discussed for indigenous groups.

Afro-Colombians have a long history of organized resistance (Villa 2001; Wade 1993b, 1995). But before the 1990s, most black groups in modern Colombia were local rather than national in orientation. In the 1970s and 1980s, urban blacks formed study and research groups, as well as small professional foundations (Rosero 1993). Among the oldest and best known of such groups was the Movimiento Nacional Cimarrón, which emerged from a radical leftist study group. In the 1970s, several students in Pereira got together to discuss problems of racial discrimination and economic marginalization among blacks in Colombia (Wade 1993b, 1995; Hurtado 1996). At some point in the early 1980s, the group chose a longtime member, Juan de Dios Mosquera, as its leader. Cimarrón's objectives were to:

—Draw attention to the discrimination and oppression of marginalized groups, especially blacks.
—Struggle for equality and universal human rights of subordinated groups all over the world, including blacks, workers, and women.
—Be in solidarity and form alliances with other black struggles, such as the antiapartheid movement in South Africa, civil rights efforts in North America, and négritude movements in the Caribbean and Francophone Africa.

Under Mosquera, and with occasional funds from international NGOs, Cimarrón gave Afro-Colombian issues a new visibility overseas. Within Colombia, Cimarrón attracted a small following among youth or cultural groups in urban centers such as Manizales, Quibdó, Cali, and Bogotá (Rosero 1993; Wade 1995). But in rural areas, where black communities traditionally organized around specific labor and social activities, Cimarrón's discourse of international black solidarity and universal human rights had little resonance. Black peasants, including women workers, were more likely to be part of cooperatives initiated by state-sponsored development programs than aligned with political activists, regardless of ethnic focus.

Since the mid-1980s, black identity and practices had become the foci of cultural revindication and celebration among various black communities but not a political force yet. For example, in 1986 the Cali-based NGO Fundación Habla/Scribe began running innovative literacy campaigns in the Pacific such as the Gente Entitada (lit., "people stained with ink") project in which participants were trained to use block-printing techniques to record and retell oral histories. According to Fundación Habla/Scribe's director, Alberto Gaona, the NGO did not engage in "adult literacy" or teach community organizing.[8] Rather, it trained community members to use various media and communication tools—producing low-cost radio programs, using videography or simple printing and publishing techniques—which the communities then used to organize themselves around issues that were important to them. Fundación Habla/Scribe's training among youth, women, students, and peasants helped these groups to organize around cultural and community issues.

During the constitutional-reform process, Afro-Colombian leaders and activists sought to unite such disparate black groups into new national-level coalitions. In Chocó, OBAPO mobilized Quibdó shanty dwellers and several coastal communities. In the southwestern Pacific, a group of students and intellectuals (many with prior links to Cimarrón) from Buenaventura, Guapi, and Tumaco began linking the struggle against socioeconomic inequalities and racial discrimination with work on revindication of black identity and cultural practices. According to Agudelo (2005), this group began in 1988 as the Fundación Litoral Siglo XXI when about thirty students and recent graduates began organizing youth and black communities in Buenaventura. Members of Siglo XXI, as it was called, linked up with the efforts being made by Fundación Habla/Scribe, other NGOs, and government entities working with black communities in the region. Inspired by ACIA's work among grassroots communities in the Chocó, these activists began mobilizing peasants, artisanal fisher folk, loggers, and miners living along the Anchicayá, Naya, and Cajambre rivers of coastal Valle del Cauca. Several activists from the Atlantic coast and Bogotá joined the group from Valle del Cauca, Cauca, and Nariño, and collectively called themselves the Organization of Black Communities (OCN). While mobilizing rural communities in the Pacific, OCN members began developing a different understanding of Afro-Colombian reality and a new vision for a national black movement within the context of the political and economic changes that were sweeping the country. The OCN's political vision was also influenced by intellectual debates about the role of culture and cultural politics in contesting development and state power, an issue I discuss in chapter 3.

Among the rural, riverine communities of the Pacific littoral there was a proliferation of new peasant associations, even as older organizations such as ACIA and ACADESAN expanded their reach. Along with ACIA, OBAPO, and Cimarrón, the OCN was a nucleus around which Afro-Colombians rallied. Even black politicians linked to the two main political parties came together for workshops and seminars to discuss the needs and interests of Afro-Colombian communities. This was an unusual development. Although political offices in the Chocó had been occupied by black and mulatto politicians since the 1950s, they had shown little interest in black resistance and protest. Rather, as Wade (1993a) describes, Chocoan politics was marked by clientelist struggles for black votes and for meager resources such as education and access to bureaucratic and political posts. At the local, municipal, and regional levels, various NGOs and the Catholic church, which had long enjoyed an important presence in the region, played key roles in rallying communities.

In the wake of constitutional reform, Afro-Colombians organized in unprecedented numbers. But there was no consensus regarding the parameters of black rights or any common strategy for framing black demands on the national stage. Chocoan proposals were couched in ethnic terms but centered on demands for land ownership and control over resources; in this way, they hoped to increase community influence in political and development circles long dominated by black elites in Quibdó. While not necessarily allies, urban groups and black politicians argued for antidiscrimination and antiracist policies to specify socioeconomic and political equality for Afro-Colombians.

In contrast to both factions, the OCN argued for a broader vision of black rights. The OCN proposal centered on two related issues:

—Respect for black difference and recognition of culturally-based economic practices of production.
—Territorial control (rather than land titles) over their Pacific homeland.

Despite these differences, the loosely aligned black factions put forward two candidates for election to the National Constituent Assembly, the body that was to frame or draft the new constitution. One was the OCN's Carlos Rosero from Buenaventura; the other was Cimarrón's Juan de

Dios Mosquera. Many other black candidates appeared on the ticket for the seventy-member Constituent Assembly. Most were linked to the Liberal Party and had not participated in grassroots mobilizations among black communities. However, during the Constituent Assembly elections of 1990, a slim majority of the black vote went to two indigenous representatives: Francisco Rojas Birry, an Emberá indian from the Chocó region, and Lorenzo Muelas, a Guambiano from the highlands of the Department of Cauca.

This was not as startling an outcome as it might appear. The Constituent Assembly was a unique legislative body in Colombia in that it was the first to include representatives from indigenous, religious, and political minorities.[9] According to Arocha (1992) and Wade (1995), the black vote was split because indigenous representatives were seen to have more political experience and, hence, a better chance of defending black interests in the framing of the new constitution.

In 1990, several national meetings were held among various black sectors, with the aim of developing and presenting a united proposal for black rights to the Constituent Assembly. At a meeting in Cali, a National Coordinator for Black Communities was formed to work on such a proposal (Arocha 1994; Grueso et al. 1998). However, the fledging Coordinator for Black Communities could not accommodate the vast differences in experience, agendas, interests, and strategies among the various black sectors and remained riddled with internal divisions. According to Agudelo (2005), ACIA and the rural contingent from the Chocó were suspicious of urban intellectual activists, and it was at the Cali meeting that the Siglo XXI group became the OCN.

Afro-Colombian activists eventually came up with proposals for ethnic and territorial rights for black communities and put them forward to the Constituent Assembly. But proposals for black rights framed in ethnic terms were met with hostility by Constituent Assembly members (Arocha 1994; Grueso et al. 1998; Wade 1995).[10] The expansion of ethnic and cultural rights was a contentious one from the outset. Many Constituent Assembly members were deeply uncertain about the granting of such rights to indigenous people. Widening the discussion to encompass Afro-Colombian demands only multiplied their doubts.

As part of the Constituent Assembly's Subcommission on Equality and Ethnic Rights, the indigenous delegates did attempt to broaden discussions about the meaning of ethnicity, culture, and national citizenship. But the subcommission's proposals were ignored, and as a consequence the term "ethnic" remained synonymous with "indigenous."[11] Subsequently, the issue of black rights was conspicuously absent in the initial roundtable discussions to draft a new constitution (Arocha 1989, 1992, 1994; Wade 1995).

Black mobilizing intensified in light of what many considered willful neglect. A massive telegraph campaign was launched to urge the President César Gaviria to support black demands: 25,000 telegraphs were sent to Gaviria urging ratification of a black law. When I visited the OCN's office in Buenaventura in June 1993, I was instantly drafted to aid in this effort. To add visibility to black struggles, activists organized several strikes and peaceful protests, including the takeover of mayors' offices in Quibdó and Pie de Pato in the Chocó, and of the Haitian embassy in Bogotá (Agudelo 2001; Arocha 1994). At local and regional levels, black groups engaged in extensive conversations about the meaning of black identity and rights. Many of the participants later became representatives of the Congress-appointed special commission for black communities which was established after 1991.

Concurrent with these efforts, discussions raged within both the subcommission and the wider Constituent Assembly regarding the ideological and practical implications of "multiethnicity" for cultural groups, other social actors, the state, and the constitutional reform process itself.[12] Especially bitter debates ensued in the subcommission over the definition of "ethnic groups" and the granting of territorial rights to Colombian blacks. Some members of the Constituent Assembly argued that only indigenous people were ethnically "different" and thus deserving of territorial, historical, and cultural rights. Afro-Colombians did not have the recognizable markers of "difference"—language, costumes, religion—that distinguished indigenous communities. Their perceived or real assimilation into mainstream Colombian society, fraught as it had always been with discrimination and resistance from white Colombians, was now held up as the very reason they could not be afforded special rights.

The momentum of black activism, however, proved impossible to stop. After much stalling, objection, and evasion, Article 7 was nonetheless included in the constitution. It declares the multiethnic and pluricultural nature of the Colombian state and allows for a broader concept of "citizen," by virtue of which ethnic groups have the right to govern themselves according to their own cultural criteria (Dover and Rappaport 1996). As a result of black lobbying, the subcommission also came up with a document aimed at doing away with the practice of equating "ethnic" with "indigenous." Amid these debates, Rojas Birry and Orlando Fals Borda consulted with various black leaders and, in conjunction with other delegates, drafted a proposal for black rights. Lorenzo Muelas presented this draft to the Constituent Assembly. At the last hour—and only when Muelas and others threatened to refrain from approving the constitution—the proposal was accepted.

Under the new constitution, indigenous groups were granted greater administrative, financial, and territorial autonomy through the proposed creation of special jurisdictional units termed "Indigenous Territorial Units" (Entidades Territoriales Indigenas; ETIs). Special rights for black communities were stipulated under the brief and ambiguously worded AT 55, which charged Congress with appointing a special commission (the Comisión Especial para las Comunidades Negras; CECN) to study the circumstances of black communities and to draft a law to grant all such communities in the tierras baldías of the Pacific zone rights to collective title to their lands. That is, the CECN was charged with drafting a law for black communities in two years, by July 1993. According to the tenets of AT 55, this proposed law would also establish mechanisms to protect black cultural identity and promote socioeconomic development among black communities in the Pacific and those living under similar conditions in other parts of the country. Nevertheless, the focus on cultural rights and socioeconomic development, and the concerns of blacks living in other parts of the country, remained secondary to the issue of clarifying property rights for black communities living in the rural Pacific.

The Long, Winding Road from AT 55 to Law 70

Law 70, the stipulated two-year successor to AT 55, was passed in August 1993—just slightly late. From 1991 to 1993, mobilization to turn the Transitory Article into law intensified at two levels: among black sectors and between black sectors and the state. In July 1992, the first National Conference of Black Communities was held in Tumaco, Nariño, in another attempt to form a broad, unified black coalition. Representatives came to Tumaco from all over the Pacific, as well as from other areas of black concentration. A stated aim of the conference was to lay the groundwork for a black law that expanded the mandate of AT 55 to recognize and include the rights of diverse black communities, not just those along the Pacific rivers.

But this is as far as the agreement went. Beyond it, divisions among the communities revealed a major rift. On one side were those who saw a law for black communities primarily as a means for obtaining extended political participation and equality within established institutions. On the other were the southwestern factions, who wanted to go beyond electoral participation and equality in state institutions to draft a proposal that built on AT 55's stipulation of traditional, collective land rights. The OCN in particular aimed to build on the momentum of black grassroots mobilization to redefine the terms of political power and economic development in the Pacific region. The OCN's aim was to propose culturally appropriate models of politics and development from an "ethno-cultural" perspective with the Pacific region as the territorial fulcrum, and the "everyday practices of black communities" providing the cultural base from which to construct these alternatives. According to Grueso and colleagues (1998: 200), the OCN wanted to emphasize the importance of "maintaining social control of territory and natural resources as a precondition for survival, re-creation, and strengthening of culture."

The dynamics within the CECN were similarly riddled with conflict. The commission got off to a late and rocky start due to government foot dragging, lack of communication, and misunderstandings—both between members and between the CECN and the wider black activist community. One challenge was interfacing with the state. Until 1991, black groups had funded the mobilization process with "soft" money from the church and various nongovernmental entities. After AT 55, there was a growing dependence on state funds, with their predictable assortment of bureaucratic strings and roadblocks. Travel funds for black CECN members were over a year in being released. A lack of basic demographic information and political maneuvers to include black politicians who had played no previous role in organizing efforts also bogged down the CECN. But the most contentious elements within it were the disagreements over how to define black identity and territorial rights. It was not until August 11, 1992, that the commission was finally established.[13]

Debates regarding black identity, reminiscent of those of the Subcommission on Equality and Ethnic Rights, raged within the CECN's Subcommittee on Ethnic and Cultural Identity. Some commission members argued that black communities had assimilated culturally and materially into Colombian society to the point of denying their African past and therefore were not a distinct ethnic group. Others suggested that Afro-Colombianness was being falsely invented as an ethnic and cultural identity based on "racial" characteristics. Arocha, the CECN's academic delegate, strongly denounced such positions and blamed them on the tendency of Colombian scholarship and society to conflate ethnicity with indigenous identity (Arocha 1994: 98). Arocha argued that the "invisibility" of black communities—despite many contributions to Colombian history, economy, and culture—was a historical wrong that needed redressing. Besides, he noted, the constitution's mandate was clear: the ethnic claims of black communities had full legal sanction.

The issue of territorial rights was equally contentious, with state officials and black representatives accusing each other of wanting strategic control over the economic and ecological resources of the Pacific. The OCN's proposal (inspired in part by the plan proposed by the ACIA–ACADESAN–OREWA discussed earlier) was to establish a collective, bi-ethnic Pacific territory under the control of indigenous and black communities. Like ACIA before it, the OCN claimed territorial rights over the Pacific, arguing that black communities had lived in the region for half a century and developed culturally distinct ways of living in, and relating to, their environment. Far from being uninhabited (*tierras baldías*), the Pacific region was the traditional homeland of black communities in Colombia. The OCN further argued that the ability of black communities to sustain their ethnically distinct and ecologically sound livelihoods depended on autonomous control over land and natural resources. They held that ethnic rights for black communities were inextricably linked to a home territory where they could "be black" and live as a community.

State officials, mindful of their own development agenda for the region, opposed such an ethnic/ecological basis for territorial control. Their proposals framed the issue either as one of property rights or in terms of collective land titles with rights to "communal" or "traditional" management of only those natural resources considered necessary to meet subsistence needs. The OCN reacted to such ideas with scorn. Its leadership claimed that property-rights proposals were being offered by the state as a way to incorporate black communities into its broader capitalist interests in the region. The OCN's mistrust of the state was greatly exacerbated in November 1992, when, despite guarantees under AT 55, the large timber company Maderas del Darién was granted logging concessions to more than 45,000 hectares of lowland forest in the Atrato region (Arocha 1994; Gómez 1994; Rosero 1993).

In the CECN, as in other fora, accommodating the heterogeneity of black identity and interests was proving difficult. A law recognizing Afro-Colombians as ethnically different found little favor among middle-class blacks, for whom such notions as collective land ownership could seem backward and discriminatory. The black position within the CECN became polarized between those who wanted to focus on getting all that was possible within the terms of existing laws and those who wanted to revise laws and redefine politics to accommodate ethnic and cultural diversity.

In late 1992, over massive black activist opposition, CODECHOCO distributed logging rights to several private interests (including the corporation Maderas del Darién) and tried to replace its regional black delegates with traditional party stalwarts. The move broke the patience of the commission members. In a short communiqué, they declared that they would resign unless the government fulfilled its obligations and sought the real participation of black communities (Arocha 1994; Wade 1995).

Meanwhile, several versions of a black law were being circulated and discussed. Some of these drafts intentionally reflected the OCN's political principles.[14] At a second National Conference of Black Communities held in Bogotá in May 1993, a draft text to be carried to the government negotiating table was revised, discussed, and approved (Grueso et al. 1998). Libia Grueso and colleagues note that Piedad Córdoba, the black senator from the Liberal Party, received a copy of this draft—the product of immense debate and toil among black activist sectors—and presented a version of it to the Colombian Congress as her own. After another three months of negotiations and foot dragging, President Gaviria signed the draft, and on August 27, 1993, Law 70 was born.

Law 70 (the Law of Black Communities) in Brief

The new law managed to expand the focus of AT 55 considerably, but land titles in the Pacific remained a central concern. Law 70 contains sixty-eight articles in eight chapters and focuses on three main issues: ethnic and cultural rights, collective land ownership, and socioeconomic development. Legally recognizing black communities as an ethnic group, the law requires the creation of mechanisms to protect their cultural identity and ethnic rights, including rights to culturally appropriate development and education—that is, to "ethno-development" and "ethno-education." It also mandates that black communities be given collective ownership of the rural, riparian lands of the Pacific coast. Collective land rights are extended to include black communities in other parts of the country "who live in conditions similar to those in the Pacific region and who use land in traditional ways." Not surprisingly, terms such as "conditions similar to the Pacific" and "traditional land use" proved quite debatable.[15] The law requires that the black communities manage these lands using "traditional" practices of production and indicates that subsistence use should precede commercial exploitation of natural resources to maintain the ecological integrity of the region. In addition, the law excludes property rights in urban areas and gives communities limited rights over renewable and non-renewable natural resources in their territories. Subsurface resource rights are particularly limited.

Law 70 also outlines opportunities, such as increased access to education, credit, and technical assistance, to help alleviate socioeconomic marginality among Afro-Colombians. It prescribes that relevant state entities assist black communities in the implementation of its mandates and that all plans and projects affecting the communities be discussed with community representatives. In addition, it requires that the National Development Council, Regional Development Corporations, and Territorial Councils include black representatives on their executive boards. The law establishes the special Division for Black Community Affairs (División de Asuntos de las Comunidades Negras) in the Ministry of Government and reserves two seats in the Chamber of Representatives of the Colombian Congress for black representatives. Finally, a High-Level Consultative Commission (Comision Consultiva de Alto Nivel; CCAN) was established to facilitate continued discussions between the state entities and black community representatives and to continue the work done by the CECN. The Consultative Commissions had branches at the department level to aid in this work. Occasional specifics notwithstanding, the terms of Law 70, like those of AT 55, are broadly conceived and subject to different interpretations.

Post-Law 70 Splits in the Black Social Movements

In October and November 1993, a third National Conference of Black Communities was held in Puerto Tejada, near Cali, to discuss the Afro-Colombian situation and establish strategies to implement Law 70

(Grueso 1993–94). As before, no agreements were reached regarding the meaning of key terms in the law or the basic tactics of the black movement. Black politicians wanted to take advantage of institutional spaces of black representation such as the Consultative Commission and the Division for Black Community Affairs. Groups from the Chocó and the southern Pacific wanted to concentrate activism at the grassroots and community levels. The fragile and unstable black coalition contained three main factions: blacks aligned with traditional political parties, organizations from the majoritarian black department of Chocó, and the OCN. It was during this effort that the OCN changed its name to the Process of the Black Communities (PCN) to emphasize the process of black struggles in Colombia, which continued even as their rights are recognized by law.

The first two groups focused their efforts on gaining entry into traditional state politics and political institutions.[16] Twelve black candidates ran for election to the two seats in the Chamber of Representatives reserved for Afro-Colombians. Most were politicians from the Pacific linked to the Conservative and Liberal parties. Among them was Agustín Valencia, a Conservative Party candidate from Valle del Cauca. Capitalizing on the political mood of the moment, Valencia adopted a rhetoric of blackness based on skin color, appropriated the slogan "social movement of the black communities," and won a seat in Congress (Grueso et al. 1998). The second seat was won by Zulia Mena of OBAPO, who attempted to articulate a grassroots black position from within the government. Mena and OBAPO also proposed an "ethno-popular" strategy to mobilize the black electorate in and around Quibdó to make a dent in elite-dominated clientelist Chocoan politics.

In March 1995, I spoke to members of OBAPO and ACIA about post-Law 70 activism in the Chocó. Given that Chocoan groups were instrumental in pre-AT 55 and Law 70 grassroots mobilizing, I was curious about how they viewed Mena's electoral position. Onni Robledo and Elizabeth Mena (Zulia Mena's sister), two young women from OBAPO, were confident that members of grassroots organizations would replace elite Chocoan *politiqueros* (mainstream politicians) in the departmental bureaucracy. Elizabeth Mena said:

> Law 70 is a big step forward for black organizations. Now their opinions are taken into account; now the community and the people are considered within the political process. Before the government used to go ahead and run projects as they pleased. Now they give the people the opportunity to voice their

opinions before the projects get under way. They ask them what will help them and what won't work.

This process [the organizing struggles during the constitutional-reform period] has also given opportunities to professionals within our community, especially to people who have worked with the process. Now there is this office, Asunto de las Comunidades Negras, headed by Pastor Murillo [a black lawyer] where one can go with concerns.

Robledo and Mena were also optimistic about these members' being able to represent the true interests of the majoritarian black electorate and bring about positive change from within state institutions. Robledo said:

This law was made for us. Each organization, each person who helped in the process feels that they helped in passing this law. They feel it is theirs, something that they did.

Also, now the Municipal Council, Departmental Council, National Planning Department, et cetera, all invite our organizations to their meetings, or at least send notices. Now they think that the black people are part of the decision-making process. This never used to happen before. Initially things just used to happen in an ordinary way with the same bureaucrats—who never took the communities into account—in charge all the time. Now the black communities have a space in the bureaucracy. They are called upon to send a representative.

We also get more media attention now. Before blacks were just seen playing sports and such. Now they show Afro-Colombians doing important things. For example, Zulia appeared on TV news, something that never happened before. Although she is part of a political party now, she participates in a different way because she comes from the organizing process of AT 55. These two posts, of which Zulia is occupying one, are a direct result of the process and represent great strides forward for us.

ACIA members were more critical of Chocoan politics and more doubtful of the possibility of black rights being addressed within existing political institutions. During our conversation, Cornelia Chavarra of ACIA noted:

In the Chocó there is only one industry—politics. They [the politicians] don't allow another way of life. They want to maintain the status quo so that the campesinos remain poor. This political industry prevents the progress of the campesinos. The traditional politicians are clientelists and liars who appropriated the black struggle and negated the work of community organizing. The government and politicians are using divide-and-rule practices, making organizations fight among themselves so they can have control over the situation.

Armed with Law 70, ACIA, ACADESAN, and other peasant associations redirected their attention from fighting logging concessions to obtaining collective land titles and developing sustainable livelihood strategies for black riverine communities in the Chocó. Farther south, the PCN's Carlos Rosero stood as a candidate for the Buenaventura Town Council elections in 1992. But Rosero lost, and after this experience, the PCN became even more critical of Colombia's bipartisan party politics (Grueso 1993–94).

As Pardo (1998) notes, the nature of the PCN in the early 1990s is hard to pinpoint: it is at once an NGO, a network of 70–120 grassroots organizations, and a sociopolitical movement. The PCN described itself as a network with three regional organizations at its core: the OCN, the Coordinating Organization of Black Communities of Cauca (Cococauca), and the Palenque Regional de Nariño. In addition to members from the Valle del Cauca, Cauca, and Nariño (the so-called VACUNA states), several Afro-Colombian groups and individuals from the Atlantic coast and Bogotá attached themselves to the PCN.

But the PCN's most visible face belonged to a group of charismatic, highly educated ex-Cimarrón intellectuals and professionals from Buenaventura, Cali, Bogotá, and the Atlantic coast. Among them were the agronomist Yellen Aguilar, the political scientist Libia Grueso, the social worker Leyla Andrea Arroyo, the historians Hernan Cortés and Alfonso Cassiani, the philosophy student Victor Guevara, and the anthropologist Carlos Rosero. Less visible in public but part of the PCN's core group were numerous women, whose vital but "invisible" roles I discuss in chapter 4.

PCN members echoed ACIA's skepticism about institutional politics. They observed that after Mena was elected to the Chamber of Representatives, she and her struggle for black rights were forced to the margins of traditional party politics (Grueso et al. 1998).[17] At a talk in 1995, looking back on a decade of organizing, Rosero remarked how the momentum for AT 55 had come overwhelmingly from grassroots riverine communities. But he noted that there was a shift away from such "organic" efforts after Law 70 passed, as black communities focused their energies on finding and sending delegates to state bureaucracies. This increasing participation also meant that black communities had to develop negotiating capacity vis-à-vis the state and other official entities. According to Rosero, the black movement was gaining traditional political power but losing its social power base. Grassroots concerns were being forgotten or subsumed within larger political economic concerns. Despite constitutional recognition of ethnic diversity, he observed, the state was not committed to giving cultural groups broad political powers. The Consultative Commission and the Division of Black Community Affairs were riddled with conflicts over who should represent black communities in these spaces, continued government neglect for black concerns, and battles over budgets and funds for the work of regional councils. Within this context, the PCN held that

> the basis of our politics is that we want the right to be black, the right to be different. Thus, while other sectors insist on institutional and electoral strategies, we put forth the idea of constructing a discourse, a language of black communities that transcends the discourse of the traditional Left, different from traditional political party lines. As a minority, or as a group differentiated socially, we seek spaces for the construction of a democracy. (Organización de Comunidades Negras 1996: 253–54)

Grueso and colleagues (1998) note that since its inception, the PCN had envisioned an Afro-Colombian movement in broader terms than the struggle to obtain legal recognition in the new constitution. During the Law 70 negotiations, the PCN's aim had been to see a broad swath of the Pacific declared black ethnic territory. The PCN was not alone in noting that the physical environment of the Pacific—ocean, rivers, estuaries, forest—had in part shaped and been shaped by Afro-Colombian symbolic and material practices. But it was the PCN that argued most forcefully that control over territory and connection to land were imperative to maintaining black identity, cultural values, and material practices. The idea of a collective black Pacific territory over which Afro-Colombians would have autonomous control of the sort granted to indigenous communities was at the center of the PCN's legal demands.

However, no interpretation of Law 70 could yield land rights to blacks equal to those of indigenous communities. And as I discuss in the next chapter, implementing such rights as did pertain to Afro-Colombians was construed by the state to mean granting legal titles within the new economic and political frameworks for the region. Consequently, at the Puerto Tejada conference, the PCN rearticulated its position in terms of the following political-organizational principles (Grueso et al. 1998: 202–3, Organización de Comunidades Negras 1996):

1. The reconstruction and reaffirmation of black identity.
2. Autonomy as blacks—that is, the right to be different and to exercise that difference by strengthening their own vision of socioeconomic development and politics.
3. Rights to an ethnic territory as a fundamental guarantee for blacks as a people.
4. Solidarity with other black movements around the world from the particularity of the Afro-Colombian situation.

Articulating such a broad ethno-cultural vision was hard enough, but making it conceptually coherent and politically viable would require time, money, and leadership. The PCN's attempts to build and implement this vision are the subject of chapter 3.

Black organizations proliferated in the post-Law 70 years. They included cultural groups; new collectives organized around logging, mining, and other productive activities; myriad community councils; and women's groups. While some of these groups had origins in earlier organizations or collectives, a considerable number emerged during the black mobilization process of the 1990s. Like the black groups of the 1970s and 1980s, these organizations had small memberships, limited mandates, and few independent funds. Many were linked rather loosely to one (or more) of the three broad factions. But they had their own specific agendas, sometimes including development projects funded by the state under the auspices of Law 70 or other initiatives. Many were nominally or actively allied with larger regional coalitions involved in implementing Law 70.

While not without its contradictions, Law 70 was undoubtedly a victory in making a space for discussion and debates about black rights. Meanwhile, as I discuss in the next chapter, the state was refining its own notions of the meaning of the law, particularly for the Pacific region, where its economic and political agenda, rather than ethnic rights, was its foremost concern.

"THE EL DORADO OF MODERN TIMES"

Economy, Ecology, and Territory

From the balcony of my hotel in Quibdó, the capital of the northwestern state of Chocó, I had a clear view of downtown and of a broad swath of the Atrato River. The plaster and cement buildings lining the malecón were cracked and molded, their once-bright paint blistered and peeling in the humid air. Next to the imposing Church of San Francisco (affectionately called San Pacho) was the town plaza—a small, cool patch of green with benches under trees. I could only just make out the faint shadow of the bust of Diego Luis Córdoba, the first black senator of Colombia. Adjacent to the plaza in little shops facing the street, merchants bought gold hand-panned from the tributaries of the Atrato. Beyond these shops and my line of sight was the alameda where I had walked earlier in the day. The stalls were full of vendors—mostly black women—selling produce and fish just unloaded off the boats, as well as myriad condiments and medicinal herbs brought from their *azoteas* [home gardens].

I turned my attention to the river, along which a ceaseless traffic of goods and people flowed. On the banks, men unloaded lumber and other wares; women washed clothes. Their calls and chatter mingled with the shrill cries of bathing children who skillfully dodged the blows directed at them by grownups when they got underfoot.

Watching the Atrato from my balcony, I recalled a view of the river from the air while flying into Quibdó. Upriver, the Atrato looked like a thick, sea-green rope. Mercury gave it this color: the mercury employed in commercial gold mining.

AUTHOR'S FIELD NOTES, FEBRUARY 1995

Even as Afro-Colombians were bringing national attention to their ethnic and territorial rights, the Pacific region was becoming a renewed target of development interventions. Government officials and the media in the 1990s variously referred to the natural-resource-rich Chocó as the "El Dorado of modern times," the "gateway to the Orient," a bridge to the economic powers of the Asia-Pacific Rim, a "tropical Eden," and a "biodiversity hot spot."[1]

Observations about the Chocó's economic and ecological wealth were usually accompanied by laments about the region's poverty and underdevelopment. According to an observer who visited Quibdó in 1970 and again in 1982, despite the rise in the world price of gold, one of the region's major exports since colonial times, the Chocó region suffered from the "inertia of underdevelopment" (Sanders 1982). Listing the numerous problems besetting the region—poor living standards, lack of basic services, high rates of illiteracy, malnutrition, birth and infant mortality—a National Planning Department (DNP) report notes that poverty "is the predominant characteristic of the Pacific" (Departamento de Planeación Nacional 1995: 2).[2] The DNP report diagnosed lack of infrastructure as one of the main causes of the region's underdevelopment: "Although the Pacific boasts the country's largest maritime port, in Buenaventura, the physical infrastructure throughout the region has not kept pace with its productive and commercial potential, which has kept the region isolated from the national economy for many years" (Departamento de Planeación Nacional 1995: 4). DNP officials and policy experts deemed that integrating "marginal" areas such as the Pacific and Amazon basins with the rest of the country was imperative to usher in a "new era of political economy" in Colombia (Departamento de Planeación Nacional 1995; Triana 1993; E. Vásquez 1994). Such economic measures were also seen as a way to modernize the economy of such "backward" and "peripheral" areas and to improve the social conditions and living standards of the local population. Accordingly, the Colombian state proposed Plan Pacífico, a macroeconomic growth and development initiative to be launched with multilateral funding. The Colombian state was exercising its "political power and economic responsibility," Thomas Sanders's (1982: 3) remedy for addressing Chocoan underdevelopment.

This development vision was not universally shared. Environmentalists observed that the projects outlined in Plan Pacífico were unsuited to the region's fragile ecology and were thus ecologically unsustainable and economically unviable (Barnes 1993). Indigenous and Afro-Colombian activists argued that the state's modernization plans violated their ethnic and territorial rights. In response to these critiques, and because of the rising visibility of environmental and cultural rights issues, the DNP redrafted Plan Pacífico to underscore "sustainable development"—economic growth that emphasizes efficient use of natural resources and environmental conservation. Colombia's participation in the 1992 United Nations Conference on Environment and Development (UNCED) also led to the inclusion of environmental and biodiversity concerns. Prominent among these was a five-year biodiversity conservation initiative called Proyecto BioPacífico, launched in 1992. Ethnic concerns also were incorporated into Plan Pacífico in the form of titling collective lands. Also included were socioeconomic development and the preservation of traditional knowledge—as components of biodiversity conservation projects.

By 1995, when I was conducting fieldwork in the Pacific, ethnic rights and environmental conservation were firmly linked to economic development in the regional rhetoric. Nevertheless, the development agenda remained macroeconomic in focus and neoliberal in orientation. Even when discussing biodiversity, Proyecto BioPacífico documents and personnel stressed its economic and strategic value over ecological concerns.

Moreover, the state and Afro-Colombian activists had very different understandings of the region's political economy and what local participation and inclusion implied. For the state, recognizing Afro-Colombian ethnic rights meant incorporating them as resource stewards into its development and conservation practices: this was the purpose, as much as the product, of land titling. For the PCN and other black activist groups, ethnic recognition implied more than property rights and participation in the state's projects. Rather, activists wished to shape the region's economic, ecological, and territorial dynamics according to their conceptions of the links between culture, ethnicity, and territory.

This chapter examines the state's economic logic and vision for the Pacific by taking a critical look at Plan Pacífico, Proyecto BioPacífico, and *Ordenamiento Territorial*, the state's territorial reorganization policy. It traces how black ethnic and territorial rights became part of these projects and discourses of sustainable economic development and assesses the Colombian state's responses to black demands. It discusses how blacks and indigenous groups counter the state's agenda by putting forth the idea of territoriality as opposed to land titles. The chapter concludes with a discussion of how notions of economic development, ecological sustainability, and ethnic rights help to establish and expand the state's power in political and cultural terms.

The Trajectory of Economic
Development in the Pacific until 1989

At various times since the Spanish occupation in the late sixteenth century, gold and other precious metals, *tagua* (vegetal ivory), timber, and fish have been the basis of the region's subsistence and exchange economies. After independence, these became key commodities in the boom-and-bust cycles of capitalist resource exploitation. The predominantly Afro-Colombian population of the region participated in and was shaped by these forces, both by direct imposition and by overdetermined "choice" (Wade 1995). Afro-Colombian communities also established parallel subsistence economies based on farming, fishing, mining, and harvesting of forest products. Although logging and mining concessions were granted to international firms, from the late nineteenth century onward a minority of mestizos and whites, and a small cadre of local black elites, controlled the regional economy and dominated politics (Agudelo 2005; Wade 1995).

Post-World War Two national development policies in Colombia followed a pattern typical of Latin America, focusing on economic growth through industrialization, urbanization, and agricultural modernization (Sheahan 1987).[3] The Chocó remained on the economic margins of the Colombian nation. Remote geography, fragile ecology, and racial demography all played their part in this isolation. But it did not long remain outside the reach of the development apparatus.

Discussions of development in the region focused on modernizing the exploitation of natural resources (Agudelo 2005; Sánchez 1995). African oil palm plantations and aquaculture operations were established in the coastal zones of Nariño and Cauca in the south. In the Chocó Department, logging, fishing, and mining intensified in the Atrato region, while banana plantations and cattle ranching expanded in Urabá.

In the mid-1950s, local entrepreneurs and industrialists successfully lobbied to establish the Autonomous Regional Corporation of Valle del Cauca (Corporación Autónoma Regional del Valle del Cauca; cvc; Escobar 1995).[4] The cvc spearheaded internationally funded megaprojects—irrigation, flood control, electricity generation, transportation, etc. While they were key factors in the capitalist transformation of the Pacific, the efforts of the CVC and other autonomous regional development corporations did not significantly improve local livelihoods. During this time, it was the Catholic church that provided many of the social and welfare functions normally provided by the state (Agudelo 2005).

In the 1960s and 1970s, a series of natural disasters (among them the Quibdó fire in 1966 and the Tumaco seaquake in 1979) brought in their wake a sharp rise in rural-to-urban migration within the region, as well as migration from the region to Andean cities. They also paved the way for state-sponsored relief and reconstruction projects. The World Bank provided loans to finance these measures; the CVC was charged with their management. Prominent among these schemes, the Integrated Development Plan for Buenaventura was designed to provide basic services and bring development to the region's population. As before, these efforts did not improve local lives to any sustained or significant degree. However, by the 1980s they had institutionalized the presence of the state and development apparatuses in the region.[5] Gustavo de Roux (1992), Jorge Ignacio del Valle (1995), Jesus Alberto Grueso and Arturo Escobar (1996), and other observers and scholars of the region argue that one reason that these development projects failed was because they were conceived without taking local sociocultural factors into consideration. However dismal the development situation, for the most part the region was free of violence and armed conflict. It is certain that small numbers of guerrillas and drug cultivators and traffickers were arriving and scattering along various parts of the coast. However, until the late 1990s they did not have a decisive impact on sociocultural or political economic life in the region.

In the broader arena of early-1980s Third World development, the "trickle-down" paradigm of economic development was under sharp critique. There were newly energized calls to pay attention to issues of "human welfare," "poverty alleviation," and "growth with equity." It was these issues and the dubious success of "top-down" approaches (those of the CVC, for example) that shaped the next set of development interventions such as President Belisario Betancur's Integrated Development Plan for the Pacific Coast (Plan de Desarrollo Integral para la Región Pacífico; PLADEICOP). With funding from UNICEF, PLADEICOP established social-welfare programs to improve nutrition and primary health care among poor communities (Agudelo 2005; Escobar 1996; Pedrosa 1996; Sánchez 1995). It also provided loans and technical assistance to small-scale farmers and fishermen to help increase their productivity. While gender was not an explicit focus of PLADEICOP, black women took unforeseen advantage of these programs (see chapter 4). A few nongovernmental organizations such as the Cali-based Foundation for Higher Education (FES) and Fundación Habla/Scribe also became active in the region: the former in funding various organizations and programs, the latter in fostering "rights-based" and "participatory" approaches to development in local communities.

In addition to social issues, emerging concerns about economic and ecological sustainability were brought to bear on third world development (Asher 2000; Lele 1991; Redclift 1987; World Commission on the Environment and Development 1987). In the Pacific, several sustainable forestry and "green" resource-management initiatives were launched to address deforestation and land degradation problems under the Forestry Action Plan for Colombia (Plan de Acción Forestal para Colombia). Among them was the Dutch-sponsored DIAR project mentioned in the previous chapter, which focused on sustainable resource use and integrated rural development along the Atrato River. In Nariño, the United Nations Development Programme funded a project called Proyecto Bosques de Guandal to study resource use in the Guandal forests of the Satinga and Saquianga river basins. (The Saquianga is a tributary of the Satinga.) In addition to sustainable forestry projects, several Colombian conservation NGOs conducted research on Pacific flora and fauna and on resource-use practices among local populations. Most prominent among these were the Bogotá-based Fundación Natura and Fundación Inguede and the Cali-based Fundación Herencia Verde. Fundación Natura conducted projects related to wildlife use and management among the Emberá indians living near Utría National Park in coastal Chocó; Fundación Inguede focused on the study of plants and the use of non-timber-forest products in black communities. Fundación Herencia Verde's projects involved research on wild and agriculture-based biodiversity in the Anchicayá River watershed of Valle del Cauca. All of these undertakings had small social and economic components, such as helping local communities secure their natural-resource-based livelihoods. But they remained largely technical projects and had little impact on the regional economy.

Despite the predominance of black communities in the project areas, culture and cultural rights did not become key components of the development discourse until the late 1980s, when struggles for black rights gained momentum. The state's presence remained marginal in most of the region; existing institutions were weak and ill funded. Co-

ordination between entiti... local, regional, and national levels
was slight; however, resourc... tion by private firms was rapidly
accelerating.

If most national institutions con...
ous or indifferent to their cause, leade...ith the Pacific were dubi-
nonetheless savvy in seeking support. The...e black movement were
cal and human-development discourses of N...drew on the ecologi-
with certain of their members. When the World ...d formed alliances
Columbian National Parks office released a proposa...fe Fund and the
serves Network (Red de Reservas Privadas), members of ... Private Re-
various Chocoan organizations approached the NGO. While t...e was ...PCN and
no consensus in the organization regarding the idea of linking b...
lands to the reserve network, the activists hoped to enlist support for
collective land claims in a sector that had not shown hostility to their
cause and might be in solidarity with it.

Plan Pacífico: A New Dimension of Economic Development for Colombia

As dismally illustrated in DNP reports, by the end of the 1980s, economic development in the Pacific (as in most parts of the Third World)
was in crisis. The blame for the failure of state-led development efforts
and for the burgeoning development-related debt was laid at the doors
of corrupt governments, inefficient bureaucracies, and imprudent social
spending. The crisis was not, of course, simply one of state-led economic
efforts. Rather, it was the end of "developmentalism" as a historical
project:

> Developmentalism was, indeed, a project originating in the stabilization of
> world capitalism after the inter-war crisis and in the context of the Cold War.
> It was a constructed order, even though planners presented development in
> ideal terms: as an evolutionary progression along a linear trajectory of mod-
> ernization. In this respect, not only would each state replicate the moder-
> nity of the First World (with the US at the apex), but there were expecta-
> tions that the development gap between First and Third Worlds would be
> progressively closed. Despite some apparent successes (e.g. the newly indus-
> trializing countries of East Asia and some Latin American states for a time),
> the development gap across these world divisions remained. The development
> project was unsuccessful in its own universalist terms. Its failure is both the

, alternative project: the globaliza-
cause and consequence of the us
tion project." (McMichael 20...by the development establishment in
This "solution" was pro...ghted "prudent macroeconomic policies,
Washington, D.C., and...free-market capitalism" (Williamson 2002).
outward orientation...d, lenders persuaded or coerced governments
All over the Thir...structural adjustment policies to liberalize or open
into the adopti...to foreign direct investment and market forces. The set
their econo...policy reforms designed for Latin America (known as the
of econo..."Wash...ton Consensus" or *apertura* in regional discourse) emphasized
fisc... discipline, export-oriented growth, and increased "public infrastructure investment" (Williamson 2002).

This logic was reflected in the vision of the DNP, which had taken over the management of Pacific development activities from the CVC. The 1989 DNP document "El Pacífico: Una nueva dimension de desarrollo económico para Colombia (The Pacific: A New Dimension of Economic Development for Colombia)" had already focused attention on the region as a strategic link to the economically powerful Asia-Pacific Rim nations. In 1992, the Gaviria administration (1990–94) followed up on the DNP's measures with "Plan Pacífico: Una nueva estrategia de desarrollo sostenible para la costa pacífica colombiana (Plan Pacífico: A New Sustainable Development Strategy for the Colombian Pacific Coast)." Plan Pacífico stressed the importance of investing in the region's infrastructure to strengthen its commercial links with the Andean interior and with foreign markets. In the last two decades of the twentieth century, several large-scale projects were proposed to modernize the Pacific economy and achieve macroeconomic goals. Diego Piedrahíta and María Estella Pineda (1993) and Bettina Ng'weno (2000) list plans to

—Construct ports at Urabá and Tribugá (in the Chocó), an inter-oceanic bridge between the Atlantic and Pacific oceans, and an inter-oceanic canal between Atrato and Truandó (also in the Chocó).

—Build several roads connecting key points in the Andean centers with regional capitals in the Pacific, among them the Popayan-López de Micay highway (Cauca); the Pasto-Tumaco highway (Nariño); the Buga-Buenaventura road (Valle del Cauca); and the Santa Cecilia-Nuquí-Bahia Solano road (Chocó).

—Install an oil pipeline from Buga to Bahía Malaga; a military base in Bahía Malaga; and hydroelectric plants in Río San Juan, Río Patía, and Arrieros del Micay (all three in the south-central parts of the Pacific).

—Award extensive commercial concessions to private firms for the exploitation of forest, natural gas and mineral resources in various parts of the Pacific.

This focus on growth had parallels with development models from the 1950s. The rhetoric of Plan Pacífico, by contrast, was the product of the more critical 1980s and 1990s. First, it paid attention to the notion of "human development," as the DNP notes:

> In the past it was believed that extending the physical infrastructure was sufficient to stimulate development in underdeveloped zones. But the experience of the past decades signals that physical infrastructure alone is not enough. Rather it has to be combined with other strategies to avoid risks and disequilibrium within the development process. For the poorest regions, human development is a basic element of economic and social advancement. (Departamento de Planeación Nacional 1992: 27)

Accordingly, in addition to investing in physical infrastructure (transport, energy, and telecommunications), Plan Pacífico included investments in social projects to improve the quality of and access to such basic services as health care, education, sanitation, and housing. Under the plan, the DNP's efforts to ameliorate the region's high indices of poverty, disease, illiteracy, and neonatal and child mortality were aided by various federal welfare programs. President César Gaviria's National Rehabilitation Plan (Plan Nacional de Rehabilitación; PNR) and President Ernesto Samper's Social Solidarity Network (Red de Solidaridad Social; RSS) are prominent examples.

Plan Pacífico also made environmental issues an explicit part of the development agenda. It declared sustainable management of natural resources and conservation of biodiversity imperative for future economic growth. Conspicuous in the draft are two projects: the Natural Resource Management Program (Programa de Manejo de Recursos Naturales; PMRN) and Proyecto BioPacífico, a biodiversity conservation initiative. Finally, in keeping with the logic of apertura and structural adjustment, the state committed to establishing both the legal basis and the institutional actors (state, NGOs, and private) involved in promoting market-led economic development. In the Pacific, this meant, among other things, clarifying property rights to reduce conflict over land and resources, establishing clear policies related to natural-resource use, and ecological zoning. Emerging from its precursor, the Forestry Action Plan for Colombia, the PMRN aimed to:

—Outline the institutional and regulatory framework for natural resource management.

—Develop sound systems for forest concessions and royalties.

—Establish policies to manage watersheds and national parks.

—Conduct ecological zoning.

—Monitor the environmental and natural resource management activities of five regional development corporations.

—Engage in land titling.

While the PMRN had both a strong ecological zoning component and a land-tenure component that focused specifically on indigenous lands (*resguardos* and ETIS), it made no mention of Afro-Colombian territorial rights. This became an issue during negotiations with the World Bank over funding for the PMRN.

In 1991, the World Bank established an "Operational Directive on Indigenous People" to address potentially adverse effects of development interventions on indigenous peoples or "culturally distinct minorities" (Ng'weno 2000). To comply with this directive, members of a Bank Appraisal mission, state officials, representatives of indigenous and black movements, and representatives of development and conservation NGOs met in 1992 in Yanaconas (near Cali) to discuss the PMRN (Ng'weno 2000; Sánchez and Roldán 2002). At the Yanaconas meeting, the PMRN and Plan Pacífico at large came under fire for paying insufficient attention to environmental sustainability and especially for ignoring the land rights of black communities. Afro-Colombian activists, then working on drafts of what was to become Law 70, were adamant that respect for their traditional economic practices and collective land titles should be an integral part of any discussions on the future of the region. As a consequence of black participation at Yanaconas, the PMRN was expanded to include the titling of Afro-Colombian lands and to involve local communities in defining appropriate forestry policies and practices. It was also agreed that regional committees consisting of representatives from indigenous and Afro-Colombian communities would be established to oversee the titling of their lands and the implementation of the PMRN (Ng'weno 2000).

In 1994, as part of its broader loan package to the Colombian government, the World Bank approved a loan worth US$39 million for the PMRN. Of this amount, US$4 million was allocated to ethnic land titling. In the same year, the Inter-American Development Bank approved a US$140 million loan for an "Agricultural Modernization" scheme to be administered by the Land and Agrarian Reform Institute (Instituto Colombiano de Reforma Agraria; INCORA). The Inter-American Development Bank also approved US$40 million to assist community groups and local government agencies in implementing elements of Plan Pacífico.

During the Samper administration (1994–98), the DNP redrafted Plan Pacífico to reflect both these changes and the 1991 Constitution's aim to make Colombia a modern nation in economic and political terms. The new version (Departamento de Planeación Nacional 1994, 1995) largely paralleled the original in its approach to underdevelopment. That is, economic growth through infrastructure and institution building remained the fulcrum of Plan Pacífico's development strategy. However, in response to ethnic activism and the increasing attention to cultural and environmental issues in global development rhetoric, the development agenda for the Pacific was expanded to incorporate the specific concerns of local communities and to increase community participation. Projects on ethno-education, collective land titles, and other elements of Law 70 became explicit parts of Plan Pacífico's social-investment portfolio. Reference to the "traditional knowledge" of local communities also began to appear in relation to biodiversity conservation. The latter had become an important strategic priority for the state, as the DNP's own publication recognizes:

> The Chocó biogeographic region possesses one of the highest biodiversity and endemism indices on the planet. Historically, the communities of the zone have exploited the resources of the region in a sustainable fashion and have managed to identify some medicinal and edible resources of the region. It is estimated that this natural richness represents a high potential for biotechnological development, and warrants further investigation and research. (Departamento de Planeación Nacional 1995: 5)

Issues of cultural rights and knowledge, economic development, and ecological concerns all came together in Proyecto BioPacífico. However, as I will discuss, these linkages were not unproblematic.

Proyecto BioPacífico: "A Dialogue of Knowledge"

Robert West, one of the first modern geographers to conduct a systematic study of the Chocó, noted in 1957 that although Europeans had exploited its mineral and forest wealth for three hundred years, the

Pacific lowlands were scientifically one of the least-known areas in Latin America (West 1957). Alwyn Gentry, one of the first biologists to alert the scientific community to the high levels of biological diversity in the Chocó, echoed this view. At a seminar in Colombia just before his death in 1993, Gentry commented that we knew more about the surface of the moon than about the biological diversity of the planet, and especially of the Chocó (Gentry 1993). Even as Gentry was making these remarks, the Chocó was becoming the locus of biodiversity research with Proyecto BioPacífico and, indeed, a key locus of concern in the new global and national environmental politics.

Proyecto BioPacífico was a five-year initiative (1993–98) funded by the United Nations Development Programme and the Global Environment Facility (GEF).[6] Its mandate was to formulate a biodiversity conservation policy for the sustainable development of the region by working with the government, nongovernmental entities, and local communities (Proyecto BioPacífico 1994). Proyecto BioPacífico had four goals:

—To recuperate, generate, and synthesize scientific and traditional (folk, ethno-biological, popular, indigenous) knowledge of Chocoan biodiversity (*Conocer*).

—To evaluate its potential economic, cultural and strategic value for local wellbeing and the common good of the nation and the globe (*Valorar*).

—To educate and organize local communities and mobilize them to participate more effectively in conservation and development activities (*Mobilizar*).

—To establish a secure legislative, administrative and financial base to protect and sustainably manage the region's natural resources (*Formular-Asignar*).

The Swiss government and the United Nations Development Programme committed US$9 million for the first phase (1993–95) to develop a participatory development strategy with biodiversity conservation as its primary goal (Casas 1994; Proyecto BioPacífico 1994). This "community-based conservation" strategy was to be drafted in consultation with local communities and specified respect for cultural knowledge, traditional resource use and practices, and ethnic land rights among its goals. The Colombian reporter Tulio Díaz (1993: 27) succinctly summarized the Proyecto BioPacífico's biodiversity conservation strategy as "the production of knowledge and the systematization of the known" through a "dialogue of understandings."

But such dialogues were fraught with conflicts. Most interested parties seemed to agree that the Chocó was rich in both "biodiversity" and "cultural diversity" and that these diversities were interrelated and needed protection. However, there was little consensus on who was to determine what constituted biodiversity and how it was to be preserved, especially in Afro-Colombian lands.

These controversies came home to me in June 1993, in Quibdó, at a Proyecto BioPacífico meeting between Afro-Colombian activists and officials from CODECHOCO, the Proyecto BioPacífico itself, and several conservation NGOs. A CODECHOCO official opened the meeting by introducing Proyecto BioPacífico's goals, drawing attention to various laws in the 1991 Constitution that mandated citizen participation. Next, two representatives from local conservation NGOs took the podium. They gave a brief overview of the biodiversity in the Chocó, how it was "the basis for all life," and why the Chocó was of national and global importance. The representatives then told the audience that Law 70 signified a celebration and recognition of Afro-Colombian cultural and ecological practices, which played a key role in managing and protecting the region's biodiversity. They concluded by remarking that the gathering was one of many opportunities for the state, black communities, and NGOs to work together to implement the ethnic, economic, and environmental policies outlined in the new constitution.

After the NGO representatives, OBAPO's Zulia Mena spoke from her seat in the audience. She thanked the speakers for explaining about biodiversity. She said that she and the black communities shared the podium's concern for the environment and were pleased that Law 70 allowed them the possibility of continuing their traditional lifestyles, which would protect both black culture and biodiverse nature. But then Mena drew a line: black communities themselves, she asserted, must be the ones to identify the diversity on their land and determine how to conserve it. She added that black communities had been living in the region since their arrival as slaves five hundred years earlier and that they and their indigenous "cousins" had devised sustainable economic practices that allowed them to live in harmony on their biodiverse lands. The participation of black communities in state plans, she emphasized, did not imply that the communities would blindly follow official directives. Rather, they would shape such directives based on their cultural knowledge. As with many such meetings, this one ended inconclusively. How black communities were to engage with conservation efforts remained unresolved.

In contrast to Mena and other black activists' emphasis on the local or cultural value of biodiversity, Fernando Casas, the director of Proyecto BioPacífico and an economist, was chiefly concerned with the scientific and economic elements of biological diversity. According to Casas (1993), implementing Proyecto BioPacífico's goals of conserving the biological richness of the Chocó—its species diversity, the "genetic bank" of its tropical forests, its extensive watershed and climatic services—would be a decisive step in meeting Colombia's national and international environmental obligations. Among the latter is the 1992 Convention on Biological Diversity, which Colombia ratified in 1994 by passing Law 165, the Biodiversity Law.[7] The Convention on Biological Diversity and Law 165 stress the conservation and sustainable use of biological diversity and the fair sharing of products made from gene stocks. These sentiments are echoed in Law 99/1993, the principal legislation governing environmental concerns in Colombia: "The biodiversity of the country, being the patrimony of the Colombian nation as well as having benefits for humankind, should be protected and used in a sustainable manner" (Law 99, art. 1, no. 2). To facilitate such protection, the Institute for the Development of Renewable Natural Resources (INDERENA) was restructured in 1993 to create the Ministry of the Environment (Ministerio de Medio Ambiente). Law 99 also established the National Environmental Network (Sistema Nacional Ambiental), which provides blueprints for conservation of the environment and the sustainable management of the nation's renewable natural resources on both public and private land. Five environmental research institutes across the country were simultaneously established to assist in the National Environmental Network's mission. Like the regional development corporations, these institutes are public entities linked to the Ministry of the Environment but with private charters and administrative autonomy.

One of the five institutes, the John von Neumann Institute for Pacific Environmental Research (Instituto de Investigaciones Ambientales del Pacífico John von Neumann; IIAP), has headquarters in Quibdó.[8] Proyecto BioPacífico and the IIAP work in close collaboration to investigate and shape biodiversity conservation policy. On January 27, 1995, the Ministry of the Environment hosted a meeting in Buenaventura to discuss the statutes that would govern the IIAP and define its functions. Over a hundred people were present at the meeting. Among them were state officials and government functionaries from the Ministry of the Environment, Proyecto BioPacífico, the CVC, and other agencies; representatives of environmental NGOs and projects such as FES, Proyecto

Bosques de Guandal, CeniPacífico, Fundación Herencia Verde, and Ecofondo; academics from the Universidad del Valle; and indigenous and Afro-Colombian activists from all over the Pacific coast.

Ernesto Guhl Nannetti, then deputy minister of the environment, presided. Guhl Nannetti opened the morning session with a speech highlighting the context, aims, and importance of the IIAP and of environmental research in general. He explained that the IIAP's chief aim was to conduct environmental research, which, like that of Proyecto BioPacífico, was to take into account the cultural, social, geographical, and biological realities of the region. Guhl Nannetti explained that the IIAP did not advocate "objective" or "blind" research. Rather, it stressed the importance of producing knowledge and "scientific truths" that could be put to use for the common good with the help of technology. Folk knowledge about the natural world and traditional ways of using and managing resources were extremely valuable within this framework, and studying traditional knowledge would be an important part of the IIAP's research. In his speech, Guhl Nannetti summarized the larger goals of SINA and the IIAP. These were also listed in the draft statute:

1. Coordinating biodiversity inventories.
2. Researching the uses and applications of biodiversity.
3. Biotechnological research and development to convert natural resources into marketable products.
4. Establishing mechanisms to recognize the property rights of indigenous and other communities to prevent the piracy of their knowledge.

The National Environmental Network, Proyecto BioPacífico, and the IIAP were part of a burgeoning set of policies, laws, and institutions governing the production and use of environmental knowledge in Colombia. All gave nods to indigenous knowledge but eventually stressed the scientific and economic aspects of Colombia's *sovereign* environmental resources and their importance in serving national goals (Asher 2001). At the IIAP meeting, Fernando Casas reiterated his view that the conservation and potential use of the Chocó's biological and genetic richness represented an increasingly important source of negotiating power for Colombia at the international level. Community and local concerns were of notably secondary interest to Guhl Nannetti.

Indigenous and black activists and their supporters saw the issue differently. Immediately after Guhl Nannetti's speech there was a motion from the floor that the agenda be modified. Yellen Aguilar of the PCN, the local schoolteacher Tresmila Rentería, and several other activists from Valle del Cauca argued that the nature and aims of the IIAP needed to be worked out in dialogue with the communities of the Pacific. Guhl Nannetti responded that the statutes to be discussed included comments by black activists at previous meetings held in November 1994, as well as black recommendations and projects proposals sent to the Ministry of the Environment. He said that the Ministry of the Environment was very interested in participation and that such participation would occur during the course of the day. The community representatives were not satisfied with this response but agreed to listen to the next speaker. Her lengthy speech on the legal aspects of the IIAP carried on to lunchtime.

The afternoon session proved no easier. It began at 3 p.m. with an argument about whether or not to discuss the IIAP statutes that had already been drafted. Opinion was divided between local activists and representatives of NGOs and other formal institutions. The activists argued that the draft statutes had been drawn without adequate input from local people who would be most affected by them. Guhl Nannetti's response, echoing other institutional representatives, was that the statutes should be discussed to assess whether they indeed were representative of local needs or not. The activists agreed to hear the proposed statutes but wished the forum to listen first to the concerns and ideas of the communities, as these were considered an important form of participation. To this Guhl Nannetti responded that the Ministry of the Environment had already proved itself participatory, as it had received about a hundred proposals, including from the indigenous organization OREWA. If the intent of this remark was to settle the controversy he was surely disappointed. One activist after another rose to call for a collective setting of the aims of the IIAP. Among them were Zulia Mena of OBAPO, Leyla Arroyo of PCN, and Orlando Pantoja of Cococauca. None of the activists were ready to settle for further discussion of the Ministry of the Environment's draft statutes. The arguments grew more heated as the hour advanced.

At four, pandemonium ensued as one official stood up suddenly and began to distribute copies of the proposed statutes. Guhl Nannetti tried unsuccessfully to bring the meeting back to order, then capitulated and declared a thirty-minute recess. During the break, the two sides huddled in anxious strategy sessions. The institutional representatives wanted to move forward to establish the IIAP as efficiently as possible; black activists wanted to take over the process so they could draft the IIAP's statutes in a way that reflected cultural knowledge and concerns. Both sides knew this was going to be a long, difficult, and expensive process.

When the meeting reconvened, Yellen Aguilar began by making a series of observations about the realities of the Pacific lowlands. First, the rhythm of the Pacific was a different one from that of Bogotá. Things moved at a slower pace. Second, he reminded the floor that the Pacific was the land of Afro-Colombian and indigenous peoples. Third, he noted that conservation and sustainable development projects in the region were managed by a series of state institutions and NGOs and not by local communities. Then he went on to present a proposal written, he declared, by the "subjects" concerned—Afro-Colombians and indigenous peoples—in conjunction with certain NGOs.

The proposal had three elements: first, workshops to be held, preferably simultaneously, in the four Pacific states. These would be spaces for black and indigenous communities to discuss their concerns about the region's status and their own. After these first meetings, there would be a collective *encuentro* (gathering) of Afro-Colombian groups. Only at a third meeting could ethnic groups, the state, and NGO representatives draft a common proposal to establish the IIAP.

Once more evoking the "rhythm of the Pacific," he predicted that the process would last through the end of February (in fact, the third meeting was held in June 1995). He told the gathering, "La participación cuesta (Participation costs)."

To the surprise of many, Guhl Nannetti accepted the proposal, but he observed at once that the government did not have funds to organize the workshops. The participants suggested that Proyecto BioPacífico provide the necessary funds. Casas flatly refused. At this point, the director of the Colombian Institute for the Development of Science and Technology (Colciencias), the federal agency responsible for promoting research in Colombia, offered to fund half the cost of the gatherings. After lengthy debate, Casas gave way: Proyecto BioPacífico would cover the remainder. Carlos Acosta of Ecofondo, an environmental NGO, also agreed to facilitate NGOs' participation in the process. When Guhl Nannetti pointed out that the black communities had to "do their bit," a voice in the audience reminded him that the communities were already contributing greatly to the process through their time, energy, and traditional knowledge.

It was also agreed that the existing Ministry of the Environment statutes would not be considered representative or valid. Before the close of the meeting, Arroyo insisted that a signed copy of the meeting's minutes be given to representatives of each organization present. She said that black activists had learned from past experience to have accords stamped and sealed; otherwise they tended to be forgotten or ignored. Finally, a committee of representatives from the PCN and OREWA, Proyecto BioPacífico, Colciencia, and Ecofondo and two independent consultants (*asesores*) was nominated to organize the next IIAP meeting.

On the following day, I attended yet another meeting, this one part of an evaluation of the first phase of Proyecto BioPacífico. The evaluation was conducted by a group of independent Colombian and international consultants.[9] In the meeting, community members and black activists from the PCN reiterated what they had said at the IIAP and other such gatherings: notwithstanding Law 70 and the claims of countless officials, local participation in conservation and development projects, as well as the cultural and territorial rights of local communities, was simply shunted. Initiatives such as Proyecto BioPacífico stressed science and economics as defined by the (non-local, non-black) representatives of the state. There was little, if any, real participation.

Marta, an activist affiliated with the PCN, recounted what she had witnessed at a Proyecto BioPacífico ethno-botany workshop. A botanist with the project invited community members, especially women, to a workshop to exchange information about local plants. The botanist passed out notebooks and pencils and asked the participants, who were mostly old women, to write down the names and characteristics of the plants and herbs they used. The botanist suggested that gathering and cataloging their knowledge about food and medicinal plants would help in efforts to conserve regional biodiversity. She also proposed that the community establish a herbarium and offered to help with the preparation of specimens and their labeling with scientific nomenclature. But participants noted that their community lacked a school and teachers and felt that establishing a herbarium was unnecessary. Many of the women could not write, were not interested in learning the Latin names of plants, and were reluctant to share their knowledge. According to Marta, Proyecto BioPacífico botanists did not understand how to handle the social dynamics of the community, tried to throw out a participant for being disruptive, and walked out of the event in tears.

Konty Bikila, another PCN activist, picked up the story and said that Proyecto BioPacífico officials had resisted the PCN's demands for more time and money to facilitate the inclusion of Afro-Colombian cultural and territorial concerns. The conservation and development agendas proceeded without them.

The assessment was widespread and consistent among members of the black community: participation and respect for ethnic rights seldom transcended the rhetorical level in the practice of Proyecto BioPacífico and other state entities. Grassroots participation in development and in biodiversity was usually restricted to informing communities of ongoing and proposed plans for the region or incorporating local people as menial workers, assistants, and informants in research projects.

Not all such critiques were external to the agencies themselves. In April 1995, I met with Robin Hissong and Mary Lucía Hurtado, two social scientists working with Proyecto BioPacífico. Hissong, a political scientist by training, began her remarks by noting that there was no public politics of conservation in Colombia and that biodiversity was a negotiation for different institutions and groups. According to Hissong, various actors in the Pacific—ethnic communities, state agencies, international donors, NGOs—came together in 1993 when Proyecto BioPacífico activities were launched.

Although its funding derives largely from international sources, Proyecto BioPacífico is designed to be implemented by the Colombian state. Lacking a clear political position or mandate, Hissong maintained, its relations with local communities are frequently contentious. In her view, the institutional gestation of the project has occurred without attention to the politics of the Pacific region. Although "participation" of an ad hoc sort does take place, she saw little evidence that either the state or the communities in question knew what they meant or wanted from the relationship.

In Hissong's account, then, the centralist tendencies of the state continue as strongly as ever. There is new rhetoric—of decentralization, participation, cultural rights—but old political practices and power structures continue, and departures from either are few. Of particular concern for her was Colombia's lack of experience with cultural diversity. While Proyecto BioPacífico collects a great deal of diagnostic information, Hissong contends, these data are not the sort that might prove useful for managing "diversity construction."

What Hissong meant by "diversity construction," and what sort of information could help manage it, she did not reveal. But both she and Hurtado observed that, for the most part, Proyecto BioPacífico's strategies reflected the approach of the state at large, which (they contended) perceives local inhabitants as a means to an end. Hurtado, an Afro-Colombiana from Timbiquí, Cauca, and co-director of Proyecto BioPacífico's Mobilizar team, also commented at length on the state's rhetorical linkage of cultural rights to conservation and sustainable development. According to her, two-thirds of Proyecto BioPacífico's budget went toward biological research. She spoke about the recent evaluation of Proyecto BioPacífico by a team of external experts and how the evaluators took little account of local critiques and dissatisfaction. Rather, the social component of the project was evaluated using quantitative measures, such as the amount of money spent on communication, the number of people who knew about the project, attendance at workshops, and so on.

Hurtado noted that the first phase of Proyecto BioPacífico, funded by the United Nations Development Programme, was supposed to take biodiversity, cultural diversity, and territorial issues into consideration. That is, the Proyecto BioPacífico was to consider the ethnic question and the issue of the communities' relations to land when outlining its conservation policies and programs. However, she maintained, the practice has been quite different. Hurtado declared that after the initial political document, the project's operative plans and other's writings have focused largely on biological diversity. Individual projects may consider traditional land use and agricultural activities, but few or none consider their social and political implications. Proyecto BioPacífico, Hurtado stated, simply decided that social movements and political processes in the region were not its business. So it has tried to implement specific projects, but with little "real" input from the grassroots. Hurtado also cited a lack of political vision with respect to ethnic groups in the region.

Hurtado concluded her interview by telling me that, in practice, "community-based conservation" activities and many other projects on traditional land use are divorced from community concerns; they become no more than "ethno-knowledge" retrieval methods. But most Proyecto BioPacífico officials overrule such objections. For the directors and technicians of such large-scale interventions, the political processes of local communities have little bearing on environmental conservation. Many officials, like Casas, perceive black ethnic and territorial demands as a strategic emulation of indigenous movements to gain access and control over land and resources in the Pacific region. Their counterparts in environmental NGOs not infrequently back them up. It was this twofold delegitimation of community critiques that the region's ethnic movements were fighting before it hardened within conservation and development practices and institutions.

Of course, these conservation efforts are closely linked to development forces. Julio Carrizosa Umaña (1993) and Juan Pablo Ruiz (1993) are skeptical about the aims of the North–South transfer of funds for Proyecto BioPacífico and note that, despite their supposed differences, developmental "geopolitics" and environmental "biopolitics" have important similarities: both are "macrospatial," and both were born from the intellectual ambitions of developed nations. Arturo Escobar (1995, 1996, 1997, 1998, 1999) locates Proyecto BioPacífico within the global biodiversity discourse and discusses at length how both phenomena represent a new "postmodern" form of "green capitalism" that ultimately reinserts both "nature" and "culture" into a system of production for material profit. These concerns were shared by black regional organizations and local communities who felt that the economic interests of the state and private capital would ultimately trump all other priorities.

Black movements were indeed trying to leverage more control over the Pacific territory by drawing on Law 70 and the tenets of the new constitution. However, the vast majority of rural black activists claimed that they desired such control not to share in the state's "political-administrative" approach to land and resource use. Rather, they claimed that their cultural logic of "territoriality" would help them conceptualize a more sustainable future for the region (Proceso de Comunidades Negras and Organización Regional Emberá–Waunana del Chocó 1995: 37).

Ordenamiento Territorial:
The Spatial and Political Reorganization of the Pacific

Among the many principles in the 1991 constitution designed to strengthen Colombia's status as modern nation was the notion of *ordenamiento territorial*, or territorial zoning, defined as "a state policy and planning instrument that allows for an appropriate political-administrative organization of the Nation, and the spatial projection of the social development, economic, environmental and cultural policies of [Colombian] society, [that] guarantees an adequate quality of life for the people and the conservation of the environment" (IGAC 2001, Definition, para. 1).[10] The Agustín Codazzi Geographic Institute (Instituto Geográfico Agustín Codazzi; IGAC) was charged with the responsibility of providing technical expertise and planning data to achieve territorial zoning for all relevant state entities. Outlining the constitutional and legal bases that underlie the politics of territorial zoning, Andrade and Amaya (1994: 36–37) note that territorial zoning helps address Article 7, which recognizes and proposes to protect the ethnic and cultural diversity of the country; Article 80, which deals with the planning, management, and use of natural resources; and Articles 103–106, which focus on achieving

a participatory democracy. Furthermore, the implementation of several
new laws depends on the development of territorial zoning. These include Law 70/1993; Law 99/1993; Law 9/1989, which addresses issues
of urban reform and zoning; and Law 136/1994, under which state administration is decentralized and municipalities are given the authority
to plan the economic, social, and environmental development of their
territories. In short, and as its definition indicates, territorial zoning is
about the reorganization of physical space and the decentralization of
bureaucratic authority. Under the terms of the 1991 Constitution and
the tenets of the Washington Consensus, both are considered necessary
to foster more efficient economic growth, greater national development,
and more effective governance.

On April 28, 1995, I spoke with Angela Andrade, subdirector of geography at IGAC, about the institute's role and projects in the Pacific.
Looking grave at the enormity of the task, Andrade remarked that un-
dertaking territorial zoning and accomplishing the state's modernization
measures required a major infrastructural and institutional overhaul of
all territorial entities (departments, municipalities, districts, ethnic territories) in areas such as the Pacific. Fernando Salazar of IGAC writes,
"the most important challenge that the region faces is harmonizing the
development of the economic and social infrastructure and the national
demand for the exploitation of natural resources with the rights of the
local communities, ecological equilibrium and biodiversity conservation
according to the principles outlined in the Constitution" (Salazar 1994:
76). I spoke with Andrade and other IGAC scientists about addressing
this challenge.[11] Perceiving it as largely a technical one, they observed
that it required basic cartographic, ecological, economic, and demographic data—data that were lacking, unreliable, or difficult to obtain.
Andrade and Manuel Amaya, the head of the territorial zoning division
at IGAC, both mentioned that existing maps for the Chocó, based on
radar and satellite images, did not provide sufficient topographic and
other details for the planning of ecological and economic activities. Obtaining accurate cartographic data using remote sensing techniques was
difficult in the Pacific because of constant and dense cloud cover. To
redress this situation, IGAC scientists began collecting and systematically organizing cartographic and ecological data in the 1990s.[12] Among
IGAC's projects were a pilot study in the Bajo Atrato region using new
global positioning systems for ecological zoning of the Pacific, a pilot
project in the lower and middle Atrato regions to measure migration
and settlement patterns; setting up better hurricane warning systems in

the coastal Pacific; and collaborating with Proyecto BioPacífico to measure and conserve various aspects of regional biodiversity (Echeverri and Salazar 1994; Gómez and Salazar 1994; Monje and Castillo 1995; Salazar 1994). Initially, social and cultural issues were not part of territorial zoning projects. However, under the terms of Law 70 and the modified PMRN, supporting the collective titling of Afro-Colombian lands and soliciting local participation in territorial restructuring projects was added to IGAC's tasks. In conversation, Andrade also mentioned that following the Yanaconas meeting, World Bank funding for the PMRN was made contingent on clarifying property rights.

Yet ambivalence about granting Afro-Colombians "ethnic rights" remained strong. Andrade, who as subdirector of IGAC had a seat on the High-Level Consultative Commission (CCAN), pointed again to the absence of clear markers of Afro-Colombian culture (e.g., language, dress), the multiplicity of black self-identifications, and the divergent ideological positions among black movements to say that "there are no real [black] ethnic communities." But while she deemed Afro-Colombian rights couched in "ethnic" terms false, Andrade believed that it was the state's responsibility to foster socioeconomic development among black communities and to ensure their welfare. Ownership of land was one aspect of such development. Guaranteeing land rights would also satisfy the terms of World Bank loans to clarify property rights in order to reduce conflicts over resources and provide legal incentives for communities to manage their lands sustainably. If the communities wanted land titles to be collectively held and maintain subsistence lifestyles, so be it. In state plans such as the PMRN, then, "ethnic territory" became primarily an issue of land tenure (Múnera 1994; Salazar 1994).

But notions of property rights, resource-use patterns, and economic practices among black communities in the Pacific are complex, flexible, and discontinuous over space and time (see the discussion of Calle Larga in chapter 3). These were being described by the Guandal Project (del Valle and Restrepo 1996) for the Satinga and Sanquinga river basins in Nariño. In March 1995, I visited the project site to get a firsthand perspective on land use and resource practices among black communities.

On March 28, I accompanied Arcelio and César, two agro-foresters with the Guandal Project, to Vereda San José on the Sanquianga River. The agronomists told me that Vereda San José consisted of forty houses with an average of six people per family. This was my first glimpse of the deep riverine areas of the Pacific, and I did not immediately realize that the *vereda* (small village) would prove to be something other than the typical cluster of homes around a church and a football field. Rather, the houses and farming plots were spread along vast stretches of the river. As in other parts of the rural Pacific, local communities engaged in a variety of activities to meet their subsistence needs, including farming, logging, fishing, and, when they could get it, paid work.

In addition to research on sustainable forestry and resource use, a small mandate of the Guandal Project included helping local communities improve their agricultural output of plantains, coconuts, cacao, and bananas. The project provided the communities with "improved" seeds and technical help. On that particular day, Arcelio was helping to lay out systematic plots for new coconut trees. Two black farmers were working in the hot afternoon sun with machetes and axes, clearing a communal plot belonging to five families. The previous year they had successfully harvested fruit from five hundred plantain and banana plants on the same plot and meant to repeat the exercise. The project also had small components to help foster community organizations and provide micro-credit loans. But as the Guandal Project was winding down, lack of funds and personnel had brought those components to a standstill.

As our boat moved along the river, I was particularly struck by the long rafts of ax-hewn and vine-lashed logs (*trozas*) being floated downriver. Young black men in tattered clothes were balanced skillfully on the logs, guiding them with long poles. According to Arcelio, logging was a seasonal activity that occurred during periods of high rain. Logging also intensified during Easter, Christmas, and other holidays or special occasions when loggers (locally known as *tuqueros*) needed to generate immediate cash for festivities or religious ceremonies. The Easter holidays were around the corner: that explained the intense logging activity we were witnessing. The tuqueros sold the timber to black middlemen, who sold it in turn to merchants in Buenaventura. Neither the tuqueros nor the middlemen earned much of a profit.

Arcelio told me about the problems caused in the region by "Canal Naranjo." The canal (*zanja*) was constructed in the 1970s by a logging firm to transport timber from the Patía River to the Satinga River. What was a small zanja soon expanded, and now the Sanquianga River swells with the water of the Patía. The results include erosion, soil damage, and other negative effects on the region's ecology and geology. Sedimentation, moreover, has destroyed local fisheries. When the swollen canal inundates agricultural lands, the communities are forced to move.

The following day we visited Vereda Naidizales—ten tiny hamlets of tuqueros living alongside a small tributary of the Sanquianga River. We stopped at one hamlet and walked for an hour through the Guandal forest, where César pointed out important tree species traditionally used by the black communities: *naidí* palm (*Euterpe oleoracea*), *sajo* (*Campnosperma panamensis*), and *machare* (*Symphonia globulifera*). When we arrived at the experimental "timber enrichment" plot, a crew of five machete-wielding men appeared and proceeded to clear vines, underbrush, and non-timber species, letting light reach the favored plants. César measured their growth. While clearing the plot, one of the men collected fruit from a wild *milpeso* palm (*Oenocarpus batana*), telling me he would later extract the oil.

Here I also witnessed the hard labor of the tuqueros, as Don Teo and Don Pacho harvested an enormous tree using nothing but their axes. Under the terms of Law 70, it was illegal to extract timber from any untitled or baldía land. Since few people had land titles, all logging activities in the region were illegal.

On our walk back to the boat, César told me about the arrival of a recent, nontraditional crop in the region: coca. While traveling along the Satinga River and its tributary the Sanquianga, he reported seeing groups of young men "dancing" on layers of plastic sheeting, crushing coca leaves into paste. The paste would then be sent to laboratories away from the river communities for further chemical purification into cocaine. In addition to farming and logging, then, coca plantations were slowly being established along the river. But César insisted that coca was still frowned on and that young men engaged in processing coca met premature death in the accelerating cycles of debt and violence sweeping the region.

The following day, I met with Luis Peña (Lucho), a Franciscan brother in Bocas de Satinga, across from the Guandal Project headquarters. Based on his ecclesiastical work with riverine communities, Peña remarked how subsistence activities and communal labor were being replaced by logging for the market. According to Peña, the increased logging meant that the relationship between tuqueros and middlemen, often one of kinship as of debt bondage, would entrench both in the new, market-oriented systems of value and exchange. He saw these changes as among the social repercussions of apertura and the crisis of the capitalism in the region. Brother Lucho believed that the black communities' struggle to obtain the rights promised in Law 70 would help them resist the negative impact of these forces.

The state's understanding of these forces and of Law 70 was, of course, different. Individual commitments varied, as did the level of seriousness with which programs and institutions approached the mandates of Law 70. But in general terms, it may be stated that the Colombian government did little in the first two years of the law to see it implemented. The reasons were several: vague terms in Law 70, unclear mandates for the multiple agencies responsible for its implementation, continued ambiguity about granting ethnic and territorial rights to black communities, and different understandings of such rights between Afro-Colombians and state officials. With respect to the first reason, multiple agencies, many of which were being restructured, were charged with addressing black rights. Three federal entities—IGAC, INCORA, and the Ministry of the Environment—were principally responsible for demarcating the collective lands and devising plans to manage natural resources within their boundaries. The task of "establishing mechanisms to protect black cultural identity" was assigned to the Colombian Institute of Anthropology and History (Instituto Colombiano de Antropología e Historia; ICANH, formerly Colombian Institute of Anthropology, ICAN), which had also been recently restructured; the government-appointed CCAN; and federal welfare programs such as the National Rehabilitation Plan (PNR), which was replaced by the Social Solidarity Network (RSS). The DNP, the quasi-private autonomous regional development corporations, and later municipal governments, were responsible for promoting "socioeconomic development according to their culture." Both the 1991 Constitution and Law 70 mandate interagency collaboration, local participation and collaboration, and decentralized forms of decision making. But all such issues (like the entire notion of cultural rights for Afro-Colombians) were without precedent and difficult to enact. Paucity of information and research on black communities and the Pacific added to the difficulties.

Lack of precedence and information notwithstanding, by the mid-1990s several projects were being launched to research black realities as part of the effort to implement Law 70. IGAC engaged a series of initiatives in collaboration with ICANH, Proyecto BioPacífico, other state entities, and NGOs to gather diagnostic information and data necessary to demarcate the boundaries of black collective lands.[13] Among them were projects to study resource use and to protect cultural practices of production in Quibdó and Nuquí in the Chocó and to outline the history of the lower Patía River basin in Nariño. There was also talk among some of the Guandal researchers to initiate a new phase of the project, which would build on their experience to help black communities obtain land titles and develop sustainable resource use plans for their lands as required by Law 70. The difficulties of unexplored methods, different perceptions of how to execute tasks and the politics of black struggles, and collaboration across agencies were evident in the Patía project.

I first learned of the Patía project on May 24, 1994, at a research colloquium led by Patricia Vargas at ICAN, which had contracted a series of studies of black communities. Vargas, an anthropologist and historian, presented the results of her pilot study on oral histories in the Chocó. As part of the study, Vargas trained community members to collect the stories and narratives of their people. The aims of Vargas's project were to obtain ethnographic information through oral histories and to enable local people to validate, develop, and revalue their own understandings of themselves.

Vargas's presentation, which also focused on the applicability of this approach in the Lower Patía Basin Initiative, generated much heated post-presentation discussion. Among the issues raised by the assembled anthropologists were questions about objectivity, the validity of data produced by research subjects themselves, and the matter of subjectivity and politics in the construction of knowledge. Also discussed were concerns over how such an approach could lead to the formation of a community "research elite," the impact of having a few people as community spokespeople, and the issue of power brokering within the community. Over the next few months, I heard rumors about funding being cut off for the Patía project. Among the reasons discussed were racism against blacks among Colombia's professional elite; unresolved theoretical tensions; and methodological debates regarding modernist, positivist-versus-postmodernist, and genealogical approaches in academic circles.

The rumors proved groundless, however, and by 1995 the Patía oral-history project was well under way.[14] The project's principal investigator, a young anthropologist named María Clara Llano, did not have much prior experience in the Pacific but had aided Zulia Mena's election campaign and was an advocate for black communities.[15] Llano headed an interdisciplinary team of three researchers: Patricia Vargas, the lawyer Miguel Vásquez, and the independent researcher Eduardo Ariza. Listed as collaborators were the anthropologist Germán Ferro of ICANH; the journalist Lavinia Fiori; and María Angulo, a community leader well known in Patía and the broader black movement. Also active on the Patía team were ten regional researchers, including teachers and leaders of the Patía Defense Council (Junta Pro-Defensa de los Patías), a community organization. Besides Vargas, many of the non-local, non-black members of the team either had experience in the Pacific or were advocates of black rights. Vásquez had drafted parts of Law 70 and had been a legal consultant to the black movement; Ariza had worked at IGAC on a document outlining rules and regulations for black land titling and collaborated with Vargas in the past; and Fiori had produced a series of radio programs on the Pacific and on black struggles.

According to the final project report (Llano n.d.), and according to Llano herself in conversation, the Patía project's aim was to construct a cultural history in close collaboration with the region's residents. Over the course of a year, the Patía team held four weeklong workshops for residents of the lower Patía basin, which includes the Patía, Old Patía, Magüí, and Nansalbí rivers. The first sessions focused on the legal aspects of black rights and the aims and methods of historical and anthropological research. Attendees were also taught basic ethnographic and oral-history methods and how to record data systematically. Subsequent workshops were open and devoted to specific themes—social cartography, recovery of oral histories, territory and settlement patterns, resource use, identity and cultural practices, and so on. A typical workshop consisted of fifteen to twenty people of different ages, sexes, statures within the community, and literacy levels, hailing from any number of surrounding veredas and rivers. Often the group broke into teams to explore subjects of particular interest.

Using their newly acquired research skills and their own ingenuity, local researchers explored their chosen themes, usually through interviews and field observations. At the next workshop, data, field notes, and maps would be shared, discussed, and revised. Research results were also shared through dramatic enactments of rituals, songs, stories, and histories. According to Llano, it was in this way that the history of the lower Patía region was collected and the final report was organized around three themes: population and settlement, territory, and identity.

Both Vargas and Llano noted that these local histories confirmed the complexity of black settlement and cultural patterns, as described in earlier works by Robert West, Norman Whitten, and Nina de Friedemann. The Patía study, like those in Chocó and the Guandal Project, also revealed the diversity and intricacies of land and resource use among black communities, who move between different economic activities—agriculture, mining, logging, fishing—depending on the availability of resources, kinship relations, accords with indigenous communities, the changing political economy, and other factors. Vargas and Llano both expressed their conviction that such research and local training would stand communities in good stead when seeking to implement aspects of Law 70, including land titling and resource-use zoning.

By Llano's account, the Patía experiment had been overwhelmingly positive, without the tense undercurrents between black communities and mestizo researchers that she had heard about in other projects. Everyone learned from each other, respecting differences and resolving disagreements with humor and grace. Beyond the research team, however, debates continued about the quality and objectivity of research conducted through such close collaboration with research subjects and how it might politicize the titling process.

There were also issues of translating nuanced ethnographic data into quantifiable or discrete forms. IGAC personnel I interviewed and whose project presentations I heard (Monje and Castillo 1995) saw land titling as an extension of zoning projects and spoke of developing and modifying geographic information systems, global positioning systems, and landscape ecology techniques to accommodate sociocultural data. They envisioned community participation in terms of "technology transfer" and training programs to help people map their lands to support their territorial claims.

To be sure, these debates reflected disciplinary differences and diverging views about participatory and applied research. But more significantly, they reflected struggles to assert or contest certain meanings. Among these were what both sides claimed as the interrelated realms of development (economic practices, property rights), nature (environment, biodiversity), and culture (ethnic identity, cultural modes of resource use, territoriality). Clearly, territorial zoning and black territorial claims were as much about defining certain parameters of resource-management practices as they were about asserting particular understandings of nature, culture, and space.

The last three concepts emerged as central at a five-day workshop in June 1995 titled "Territorio, Etnia, Cultura e Investigación en el Pacífico Colombiano (Territory, Ethnicity, Culture, and Research in the Colombian Pacific)" held in Perico Negro, Cauca. The specific aim of the workshop, convened as a follow-up to the January IIAP meeting, was for black and indigenous representatives to draft their proposals for statutes for the research and activities to be conducted by the IIAP and to lay the groundwork for the collective land-titling process. But as the preamble to the published proceedings notes: "Yes, we outlined a proposal of statutes. But . . . our aim is to strengthen our organizational processes, which will allow those of us who live in the Pacific to come up with a proposal that will defend our Pacific and through that defend life for all" (Proceso de Comunidades Negras and Organización Regional Emberá–Waunana del Chocó 1995: 14. Hereafter PCN–OREWA). The Perico Negro meeting was an unprecedented historical event in that organized indigenous and black groups gathered together for the first time since the new constitution had been adopted to discuss their mutual concerns and positions vis-à-vis plans for "their" Pacific.[16] Forty-eight representatives from thirty-three ethnic groups (key among them the PCN, OBAPO, and OREWA) were present at Perico Negro. Thirty representatives from state agencies and NGOs were also present as "guests" to bear witness to and support the efforts of ethnic activists.

For two days I was among the guests at Perico Negro. Unlike at the IIAP meeting in Buenaventura, there was little emphasis on efficiency, brevity, or decision making. The abundant debates, information sharing, and storytelling were recorded by Fundación Habla/Scribe and published later that year (PCN–OREWA 1995). The document outlines in rich detail the discussions, grievances, and political positions voiced by black and indigenous groups at Perico Negro. Here I draw attention to two elements of these deliberations: a set of principles perceived as orienting interethnic relations, and the indigenous and black "cosmovisions" (understandings of their lives in relation to other humans and non-humans) and concepts of "territoriality." The groups claimed that these perceptions guided their analyses of the region's problems and were the framing principles of their proposals for the role of science, technology, and traditional knowledge in the IIAP.

The ethnic groups present at Perico Negro agreed that relations between indigenous and black communities should be based on the following four declarations and that these same principles should guide their relations with other social sectors (the state, NGOs, and private entities):

1. The geographic area of the Colombian Pacific is the ancestral territory of culturally diverse ethnic groups.
2. The ethnic groups of the Colombian Pacific include black communities in all our diversity, and distinct indigenous peoples: Tules, Katios, Wounaan, Eperara Siapidara, Chami, Awa, Zenúes, and Emberás, who affirm the right to acknowledge and respect the differences among ourselves and from the rest of Colombian society.

3. With mutual respect and tolerance as fundamental principles of our co-existence and our strength, we, the black and indigenous peoples, declare that we are in solidarity in our struggle to defend our ancestral Pacific territory.

4. Our traditional knowledge and understanding are the basis of life, of our relations with nature, and are essential elements of our identity as ethnic groups. Indeed, they are our cultural patrimony and should be so acknowledged and respected in all their diverse forms at national and international levels. (PCN–OREWA 1995: 17)

The list ends by noting that "it is the obligation of the Colombian state to protect, respect, acknowledge and value the collective knowledge of the said peoples and communities." In contrast to state proposals, these principles give primacy to "ancestral" (as opposed to national) claims over Pacific territory and emphasize cultural diversity and traditional knowledge over economic relations and resource-based development. That culture and ethnic relations are considered linked to the political economy is evident from the implicit reference to traditional modes of economic production ("relations with nature") and the explicit one to state power in the fourth declaration.

Black and indigenous cosmovisions follow closely from these principles and reiterate that the link between cultural practices, identity, and territory is fundamental. Both groups differentiate between land rights and the notion of territoriality, noting that only the latter encompasses the idea of territory as a social, cultural, and material space. For indigenous peoples, conceptions of territory often corresponded with origin stories and myths that shape the material and spiritual life of communities (PCN–OREWA 1995: 20–29). Among black communities, notions of territory are variously linked to and shaped by historical settlement patterns, cultural conceptions of nature, production practices, and the relationship between humans and their environment (PCN–OREWA 1995: 30–37).

At Perico Negro, both indigenous and black groups expressed concern about the exploitation of the region's natural resources, noting that unregulated extraction often led to violent conflicts that threatened their material and cultural existence. Neither group perceived capitalist macro-development and linkages to world markets as providing solutions to such conflicts. Afro-Colombian activists observed that, rather than ameliorating the situation, economic growth led by the private sector would bring in drug traffickers eager to find new ways to launder illegal money. Other armed actors would inevitably follow, as would escalated levels of violent conflict—as, indeed, happened a few years later.

Proyecto BioPacífico and the Ministry of the Environment's biodiversity-conservation efforts were also criticized as extensions of economic forces that would ultimately compromise environmental concerns and community interests. While acknowledging that many within the state and NGO community shared their concerns, Afro-Colombian activists rejected the latter's faith that "institutional" approaches, with their emphasis on economic growth and Western science, were the path to real "sustainable development." Echoing the arguments that they made in the Special Commission for Black Communities while negotiating the draft of Law 70, black activists claimed that alternatives to economic and ecological problems would only emerge from the cultural logic and organizational processes of ethnic communities (PCN–OREWA 1995: 48–54).

The Perico Negro declarations provided grounding for the subsequent ethnic proposals for IIAP research statutes. Both the Ministry of the Environment and Perico Negro draft statutes are legal charters written in technical language and are similar in many respects.[17] What distinguishes them most is in the degree to which local participation and perceptions would shape the research and actions of the IIAP. Whereas the Ministry of the Environment draft sees ethnic concerns as important but subsidiary to the applied ecological research agenda, the Perico Negro draft considers these as guiding or framing elements of the IIAP. For example, the Perico Negro draft decrees that there be four, as opposed to two, representatives of ethnic communities in the IIAP's board of directors and that they be involved in discussing research projects rather than just informed of an agenda set by others. The Perico Negro draft also urges that traditional knowledge and local practices related to the environment be validated on par with the technical and scientific information produced by scientists and other professionals. Overall, the Perico Negro draft emphasizes the need for more space for the voices and participation of ethnic groups within the IIAP.

At the end of the Perico Negro meeting, representatives of all three sectors—ethnic communities, the state, and NGOs—agreed to continue dialogue before reaching a final decision on the draft statutes. But the Ministry of the Environment, as the chief representative of the state in these negotiations, continued to hedge in committing funds or setting dates for such meetings.

Despite the foot dragging, the victories at both IIAP meetings were significant. As Aguilar's comments at the January meeting foresaw, the process of engaging black concerns and participation was complex and slow. Negotiations over the IIAP continued for almost four years. The final statutes adopted in May 1999 did have far more room for the cultural concerns and rights of local communities than the original drafts.

For Afro-Colombian activists at Perico Negro, especially the PCN, the IIAP and conversations with state and non-state sectors were part of broader struggles to obtain autonomous control over the Pacific territory to redefine the regional political economy on ethno-cultural principles. They conceded that this requires black groups to be "bilingual"—able to communicate between the "modern scientific" discourse of the state and other sectors, and the "rural scientific" language of the communities (PCN–OREWA 1995: 37). As the next chapter describes, such bilingualism was a logistical and ideological challenge for the as yet nascent and divided black movement, and particularly for the PCN.

For its part, the state's intentions and ability to grant ethnic groups greater autonomy and territorial control remained open to question. Many professionals affiliated with or contracted by state institutions were strong advocates of black rights. Many expressed as much concern about economic globalization and apertura as did the ethnic communities. But for the most part, they were either unwilling or unable to work beyond the structural constraints of the political economic conjunctures. Following lengthy exchanges with black movements, the PMRN, Proyecto BioPacífico, territorial zoning, and other projects did refocus somewhat to accommodate cultural differences and ethnic rights but largely within the binding logic of economic development.

The PMRN specifically focused on establishing regional committees and community councils to work on demarcating collective black lands. But as in the past, it continued to experience difficulties—inadequate funding, lack of clarity about the criteria and methods for demarcation of collective lands, conflicts over land among and between ethnic groups, weak regional committees, lack of communications between institutional officials and between the various institutions involved, and problems with granting titles in ecologically sensitive zones such as mangroves (Sánchez and Roldán 2002). After 1996, the forced displacement of massive numbers of people from certain rivers by paramilitary and other, unidentified armed groups became a major humanitarian crisis and posed setbacks even for land titling. By the end of 1996, 13,000 people had been displaced. Despite these problems, Sánchez and Roldán (2002) note that fifty-eight collective land titles covering 2.3 million hectares of land were given out between 1995 and 1999.

Land titling was part of a broader territorial zoning effort to demarcate both physical boundaries and the discourses within which to understand and produce the new realities of the Pacific. If success were measured by the diffusion of terminology alone, territorial zoning would be a triumph: people throughout the region began to speak and write of "biodiversity," "ecological zones," "sustainable development," "property rights," and "administration and governance." Territorial zoning projects were also effective in organizing these new realities into distinct territorial units—parks and reserves for biodiversity, management zones for forests and watersheds, collective lands for black communities. The state and its subsidiaries began the tasks of managing, governing, and establishing decentralized jurisdiction over these territorial entities. Local communities trained by specialized territorial zoning personnel also began to use the logic of these entities and the new parameters for their governance.

Economy, Ecology, Ethnicity, and the Production of the State

At the end of the twentieth century, the Colombian state was both strong and weak. It was strong, because it excluded large parts of the population from its oligarchic democratic politics. It was weak, because its legitimacy and power were challenged from so many sides—guerrillas, drug traffickers, counterinsurgent paramilitary forces, and social movements of various stripes. With its highly centralized bureaucratic structures, the state had little presence in or control over so-called marginal regions such as the Amazon and the Pacific coastal areas. However characterized, in the period leading up to the constitutional reform, Colombia's credentials as a "modern" nation-state were questionable. After all, a modern nation-state, at the minimum, is expected to have jurisdiction over a defined territory and enjoy the support of its citizens.

It was partially to address this issue of dubious state legitimacy that Colombia adopted a new constitution in 1991:

> [These] are the essential aims of the State: serve the community, promote general prosperity and guarantee the effectiveness of the principles, rights and responsibilities written in the Constitution; facilitate the participation of all in

the decisions that affect them and the economic, political, administrative and cultural life of the Nation; defend national independence, maintain the territorial integrity and assure the peaceful existence and . . . of a just order.

The Republic is granted authority in order to protect all residents of Colombia, their life, honor, property, beliefs and other rights and liberties, and to assure that the State and private individuals comply (with?) their social responsibilities. (Article 2, Colombian Constitution 1991)[18]

The 1991 Constitution also reflected a moment of promise and hope that the Colombian state would move beyond its problems—oligarchic democracy, contradictory economic policies, low levels of political participation, uneven attempts to address social inequality and tensions, environmental problems–and actually implement some radical social and political changes. The current crisis and recent commentaries such as those by Ana María Bejarano (2001) indicate that most of those hopes were not realized. Yet I argue that, despite its failure to fulfill many of its constitutional promises, it successfully set into motion certain processes and ideas about politics, economics, and rights that resulted in an increase in the presence and influence of the state in the Pacific. Not only were the new economic and ethnic discourses in the Pacific the idiom through which economic-development projects unfolded, but they had the contradictory effect of establishing more firmly the presence of the Colombian state through the expansion of its bureaucratic apparatus and its legal discourse.

In keeping with the economic logic of globalization, Plan Pacífico's central focus was on developing the resource-rich Pacific region and linking its economy with regional and global markets. Unlike in past development initiatives, the state's role was to set up policies and legislative frameworks to facilitate growth through market forces rather than to administer or spearhead economic enterprises. Decentralizing bureaucratic authority and strengthening "civil society and institutions" were also part of the new political economic rationale. But as James Ferguson (1994) notes about a decentralization effort in another space and time, despite its "de-politicizing function," this avatar of development also had "powerful political implications."

New policy instruments such as the PMRN and territorial zoning helped establish a state presence and the idea of the state as a central and legitimate focus of polity. These instruments, as well as programs such as Plan Pacífico and Proyecto BioPacífico, mobilized a certain set of discourses, a select group of ideas and structures through which life and territory in the Pacific was to be imagined. These projects (re)imagine or (re)fashion "marginal" and "backward" areas of the Pacific to integrate them not only into markets but into a state apparatus, as well. Proyecto BioPacífico and territorial zoning techniques map and organize various elements within state boundaries: its ethnically diverse citizens, its bio-diversity, its natural resources that are to be managed or exploited.

The naming and organizing of these elements also function to pro-duce the idea of the Colombian state as the guarantor of rights, the manager and steward of economic and ecological resources. That is, not only do Proyecto BioPacífico and territorial zoning catalog the people and resources of the Pacific, but they also declare and establish the *very criteria* through which resources should be developed, biodiversity con-served, and local needs met. Various scholars and activists reflecting on such discourses (Escobar 1997, 1999; Grueso et al. 1998; López 1999; Peet and Watts 1996; Shiva 1990) assert that biodiversity conservation strategies are part of a multilayered process by which the Colombian state attempts to assert its presence and control in the Pacific region, even as it inserts the material resources of the region into global circuits of capitalism.

The new policies and practices of development and conservation bring to mind Michel Foucault's notions of biopolitics or biopower—the gov-erning of subjects and populations in national space (Foucault 1980). In his research on the "question of government" and governmentality, among other things, Foucault traces the constitution of political econ-omy, which, he notes, "arises out of the perception of new networks of the continuous and multiple relations between population, territory and wealth" (Foucault 1991: 101). Colin Gordon extends the observa-tion, pointing out that the linkages between biopower/biopolitics and the theme of government allows us to trace how "the terms of govern-mental practice can be turned around into focuses of resistance: or as he [Foucault] notes in his 1978 lectures, the way the history of government as the 'conduct of conduct' is interwoven with the history of dissenting 'counter-conducts'" (Foucault 1991: 5).

The way black and indigenous communities countered the implemen-tation of instruments such as territorial zoning, the Proyecto BioPací-fico, and the PMRN are a case in point. But one should not overestate the utility of Foucault's model: as Stuart Hall (1988: 53) observes, it is a conception of power without a conception of hegemony.

The latter is an oft-cited and deceptively simple concept. For the economically non-determinist, historically and locationally specific theorist Antonio Gramsci, hegemony is always relational and never complete (Gramsci 1995 [1971]). Gramscian hegemony is similar to Foucault's micropolitics of power in the sense that people accept and spontaneously consent to the modes of domination prevalent in society. But Gramsci's complex formulations of state–society relations, wherein the state is both coercive and formative, allow us to see the particular ways in which the state, as a specific historical bloc, exercises ruling power through a combination of coercion and consent among other groups. Foucault's insights on biopolitics and the "practices" (rather than institutions) of government and Gramscian notions of hegemony bear on my critical readings of state policies and black resistance in the Pacific.

The process of state formation through a discourse of modernization and development in marginal areas is neither new (indeed, it has existed since the nineteenth century) nor without contradictions. State formation is helped by the very issues and entities that were ignored in the state's initial economic formulation. The inclusion of ethnic concerns in legal, institutional charters, the demarcation of collective lands, the participation of local communities, the gathering of ecological and cultural data—all were meant to help decentralize state power and functions. It was often ethnic and local groups who demanded that such efforts be recognized and valued. But these very activities set into motion processes that accelerate and expand bureaucratic state authority (Ferguson 1994). As I discuss in the next chapter, black movements in the region are attempting to construct cultural alternatives to contest state authority and the development apparatus while simultaneously being shaped by these forces.

Views of the Río Atrato from the air.
All photos by author.

Produce on the Río Guapi.

Measuring tree growth in a
Guandal forest in Nariño.

Tuqueros (loggers) in a Guandal forest in Nariño.

Trozas (logs) lashed together with vines being transported in the Sanquianga River basin, Nariño.

March against Racism along Calle Quinta
(Fifth Street), Cali, March 21, 1995.

March against Racism, Cali,
March 21, 1995.

Don Miguel
hand-hewing
a canoe.

Doña Naty
participating
in a mapping
exercise in Calle
Larga.

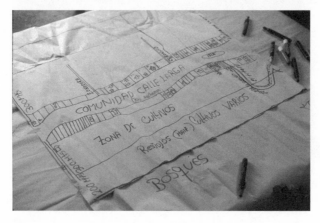

Participatory
map of
Calle Larga,
Anchicayá
River.

March on International Women's Day,
Quibdó, March 5, 1995.

The women of CoopMujeres,
Guapi, Cauca.

✻ 3

"EL RUIDO INTERNO

DE COMUNIDADES NEGRAS"

The Ethno-Cultural Politics of the PCN

After Law 70 was passed in 1993, the fragile coalition of black movements that had drafted and negotiated the black law split apart. Ensuring that black rights were granted by implementing the various tenets of
Law 70 became the de facto focus of most post-1993 Afro-Colombian
movements. But, as noted earlier, the terms of Law 70 were broad and
imprecise and subject to conflicting interpretations.

Two of the central objectives of Law 70 are to recognize and protect
ethnic diversity and to ensure that black communities participate in decisions that affect them. Chapter II of Law 70 delineates its principles:

> —Recognition and protection of the ethnic and cultural diversity and the
> right to equality of all the cultures of the nation.
> —Respect for the integrity and dignity of the cultural life of black com-
> munities.
> —Participation of the communities and their organizations, without detri-
> ment to their autonomy, in all decisions that affect them.
> —Environmental protection, focusing on the relations between black com-
> munities and nature.

On the basis of these principles, black politicians linked to the Conservative and Liberal parties, and to a lesser extent groups from the Chocó and
movements such as the Movimiento Nacional Cimarrón, turned to traditional mechanisms of political participation to gain social, economic,

and political equality for black communities. In contrast, the Process of the Black Communities (PCN) mistrusted the state and its intentions and took an anti-institutional stance vis-à-vis Afro-Colombian struggles. In January 1994, in an interview with Arturo Escobar, PCN leaders noted, "The state wants to close the door and leave the window half open so that not many people can pass through, nor much air. . . . For the state, Law 70 is a way of giving legal rights to the blacks, of institutionalizing their problems and concerns. This political opening is "a cushion of air" for the economic opening. So if the problems are institutionalized they can be managed (Organización de Comunidades Negras 1996: 247). Within this context, PCN activists noted:

> The black communities of Colombia need *espacios propios*—rights over our territory—but also spaces where we can consolidate our positions. We have differences within, we have a different position than Cimarrón, . . . or the politicians. Our purpose is to form and to strengthen the process of organization so that the very same communities generate processes of resistance or alternative visions to confront what is being imposed on them by the state.
>
> We aspire to construct a new society, where we all fit with all our differences, where no one is excluded or marginalized or segregated, where there is mutual respect for differences. We are not against the concrete demands of water, electricity, health care, transport. But we had to take a position—respond to a series of concrete petitions or work to open up spaces from which we can formulate our politics. We chose the latter. (Organización de Comunidades Negras 1996: 252)

Obtaining espacios propios—conceptualized variously as autonomous physical territory, political space, and ethnic homeland—was central to the PCN's understanding of ethnic rights. During numerous conversations, in public fora such as IIAP meetings and in their talks and writings, PCN leaders and activists reiterated such opinions. Drawing on the terms of Law 70, the PCN aimed to organize a black social movement according to the politico-organizational principles of "identity, territory, and autonomy" outlined at the third National Conference of Black Communities, held in 1993 in Puerto Tejada. The PCN's key goals were:

> —To organize a broad, grassroots-based black social movement based on diverse Afro-Colombian identities and interests.
> —To envision a political strategy that would enable organized black communities to make autonomous decisions regarding their livelihoods.

—To develop culturally-appropriate, ecologically-sustainable models of economic development.

—To establish autonomous territorial control over the Pacific.

Like the terms of Law 70, the PCN's goals were lacking in specifics and ambitious in scope.[1] In the short term, its strategies for organizing a broad-based black movement as an alternative to state and development hegemony had limited success. Numerous factors contributed to this outcome.

First and foremost, black interests were far from uniform and often did not conform to the group's baseline declarations concerning black identity and "difference." "Black culture" is composed of diverse traditions and flexible practices. Capturing this diversity and heterogeneity under one model of "culturally appropriate development" poses cultural and political dilemmas, to which the PCN did not have sufficient answers.

Another problem was largely logistical. Consensus building and the construction of viable strategies at local, regional, and national levels were tall orders for a nascent movement like the PCN. Even for organizations with deeper pockets (e.g., established political parties) and longer histories (e.g., Cimarrón), mobilizing at the grassroots is among the hardest and least certain of undertakings. The situation was made even more difficult by the PCN's limited reach in rural areas. Indeed, it could be said that the black struggle itself scarcely existed for the most isolated communities, who were frequently unaware of Law 70's existence and saw little of themselves reflected in visitors or proposals of the PCN. This was a fundamental contradiction of a movement that saw itself as emerging from, and representing, the needs of rural black communities.

Neither PCN leaders nor rank-and-file activists had much experience organizing across the differences within black communities. This deficit handicapped both its efforts to mobilize communities and to make demands of the wider nation-state—especially given that the state was largely unsympathetic to these demands. Furthermore, while official antagonism towards the notion of black "autonomy" certainly limited the scope of possible alliances, the PCN was undeniably constrained by their unwillingness to join forces with NGOs and other sectors sympathetic to black struggles. Lack of experience in dealing with complex bureaucratic institutions also hindered the group's work.

Autonomy as a goal, moreover, was at the very least problematic. Black communities could not simultaneously divorce themselves from state,

NGO, and development interests *and* negotiate with them over the region's future. This contradiction becomes especially clear when one considers that the PCN often had to depend on funds from these entities to mobilize its base. Indeed, the very meaning of autonomy is far from clear given the complex relations between black social movements and the state.

At a talk at UniValle in August 1995, Carlos Rosero said that government officials often referred to the process (*el proceso*), or organizational dynamics, of the black movements as *"el ruido interno de comunidades negras* (the internal noise of black communities)." In this chapter, I examine this "noisy" process of black organizing. I begin by highlighting the diversity of perceptions, voices, and political outlooks represented within black movements. Next, I profile the PCN's efforts to draft a culturally appropriate model of economic development for the Pacific according to its ethno-cultural principles. I discuss the complex norms of land and resource use at the local level and how these (among other factors) hampered the formation of such a model. I conclude by discussing how the heterogeneity of black identities and interests cannot be understood through such binaries as "equality versus difference" and caution as much against romanticizing resistance as against reducing social struggles to structural effects.

Those with Identity and Those without:
Organizing and Political Autonomy

According to Articles 48 and 56 of Chapter VI (Mechanisms to Protect Cultural Diversity) of Law 70, all regional development corporations in the Pacific region must include an elected representative of the black communities on their boards of directors. To comply with this requirement, the board of directors of the Autonomous Regional Corporation of Valle del Cauca (Corporación Autonóma Regional del Valle del Cauca; CVC) convened a public meeting of black communities of the department on February 4, 1995, at their headquarters in Cali. A representative of the black communities was to be elected at that meeting. In the early afternoon of February 4, a large crowd of representatives of black organizations waited in the lobby of the CVC building while board members and state coordinators met in private to discuss formats and protocols for the election. Those in the lobby included members of various Cali groups (including a local chapter of Cimarrón), PCN representatives from the coastal town of Buenaventura and the surrounding rural areas, and black politicians affiliated with the regional and departmental branches of the Conservative and Liberal parties. A significant number of young people and women were present at the meeting.

As the minutes turned to hours, the waiting crowd moved from the lobby into the large assembly hall where the meeting was to be held. They began a steady rhythm of songs, chants, dances, and speeches celebrating black identity and liberty. Someone pinned a flag featuring the names of several PCN organizations on the podium between the flags of the nation and the Department of Valle del Cauca. The flag was painted in yellow, black, and green—an intentional nod to South Africa's African National Congress. A competition of sorts ensued between the various organizations. A group of young musicians from Aguablanca (the shanty town near Cali where many blacks live) calling themselves Grupo Ashanty performed rap, reggae, and hip-hop. Members of the coastal organizations entertained the crowd with *currulaos*, *alabaos*, and other expressions of black oral tradition to the sound of *bombos*, *cununos*, and *guasás* (traditional musical instruments from the region) they had brought with them. Black politicians in dark business suits who were not in the private meeting watched these activities but did not participate in them.

After three hours, at 4:30 p.m., the state coordinators adjourned their closed-door conference. Joining the crowd assembled in the hall, they proposed to call the meeting to order with a rendition of the Colombian national anthem, followed by the anthem of the Valle del Cauca. While some members of the black groups joined in the singing of the anthem, others interpolated the singing with their rendition of the *Himno de los Negros* (Anthem of the Blacks). The government officials sat down after they finished singing the department anthem, well before the black anthem was finished. The director of the CVC proceeded with a speech and outlined the meeting's agenda.

Several members of the gathered black communities began protesting these proceedings. They said that the CVC members had been disrespectful of black culture by sitting down in the middle of the rendition of their anthem. They furthered demanded that the communities have a say in setting the agenda of the meeting. The officials on the podium ignored these demands and continued with the speeches while the crowd began to mutter in protest. After the fourth speaker, the heckling from the crowd made it impossible to hear the speech. At this point, Pastor Murillo, a black lawyer from the Division for Black Community Affairs, who had just arrived from Bogotá, stepped up to the podium and took the microphone. He implored the crowd to calm down and asked to hear their demands. It turned out that they were divided: some groups wanted to continue with the CVC agenda and proceed with the election of a black representative to the board. The groups linked to the PCN demanded that the agenda for the meeting and the election rules be determined in an open forum, beginning with a discussion of the antecedents to the meeting. CVC officials argued that these discussions had already been held and that talking about them in an open forum would delay the election. They proposed to move forward with the election, which was one of the main goals of the meeting. A CVC official invited candidates to come forward and give their speeches. He said ballots would be distributed after the speeches concluded.

Again, PCN members objected, citing a prior discussion between CVC officials and black leaders where it was decided that a representative to the CVC board would be elected through an organizational vote. PCN members wanted each organization, rather than each individual, to cast a ballot, asserting that such a process reflected the traditional Afro-Colombian logic of making decisions collectively. CVC officials denied that any such agreement had been reached.

The dispute lasted late into the evening. PCN members complained that the election was being hijacked by politicians and accused them of bribing Cali groups for votes. Pastor Murillo tried to negotiate between the two factions, reminding the crowd of the history of black struggle and asking them to behave in a manner that would uphold the dignity and unity of all Afro-Colombians. The PCN faction applauded and agreed with Murillo, while several others shouted that the way to uphold black dignity was to participate calmly with the meeting and election at hand. At about 10 p.m., PCN leaders declared the election invalid for violating prior procedural agreements and withdrew their candidate from the race. Non-PCN groups did participate in the election, and one of the two Cali-based candidates was elected as the black representative to the CVC board. This candidate was allegedly linked to Agustín Valencia, a black politician from Valle del Cauca. Valencia, then a member of the Conservative Party, was an elected member of the Chamber of Representatives.

The dynamics of the CVC meeting were not exceptional. In such *espacios mixtos* (mixed institutional spaces), a state official would call the assembly to order and open with a rendition of the Colombian national anthem. PCN members would either listen politely and then follow the

Colombian national anthem with the *Himno de los Negros* or interrupt
the former with the Black Anthem. Next, the stony-faced state officials
would present the agenda of the meeting to the group. Once again, the
official statements were heard politely, then PCN members would suggest that it was time to set a common agenda in a participatory fashion—
one in which everyone in the room would get a chance to voice his or her
views. PCN members saw this as a way to set an agenda collectively and
with a focus on oral communication. Everyone could speak, even if the
speakers reiterated the same points. Indeed, in informal PCN meetings,
as well as in formal espacios propios (spaces for dialogue among black
groups alone), each point of discussion was subjected to much scrutiny
before the discussion moved to the next topic, often after several hours.
Discussions frequently ended only when people packed up for the day
(or, frequently, the night) without having reached any decisions.

State officials witnessing or engaging in these extended dialogues were
often impatient or frustrated with the PCN's strategy. Some viewed it
as disruptive and disrespectful behavior by "ignorant blacks who do
not know how to organize efficiently and effectively." PCN members
considered such lengthy discussions an inevitable part of the "growing
pains" of the black organizational process. They asserted that participation meant having "a voice, rather than just a vote" on issues selected
in advance. According to Álvaro Pedrosa, a professor of communications at the Universidad del Valle who has studied black oral traditions,
among local Afro-Colombian communities, the purpose of discussions
is a "point–counterpoint" exchange rather than "progressing" with the
theme or the issue at hand (Pedrosa et al. 1994). That is, engaging in
polemics is a central feature of social and political encounters. Like Yellen Aguilar at the IIAP meeting, Rosero observes that local communities have a very different sense of time and political process than that
of state bureaucrats. Maria Lucia Hurtado (1996) also notes that black
organizational strategies differ from national and Western norms. She
contends that constructive dialogues between state officials and black
groups are difficult because local cultural differences are undervalued or
misunderstood.

Understanding local differences within the movement was also a chal-
lenging task. On the day following the CVC incident, I was talking to
Mercedes Segura of Fundemujer, a women's organization based in Buenaventura, who had attended the meeting as part of the PCN delegation. Reflecting on the dynamics of the meeting and the various black
delegates, Segura distinguished between people "with identity (*con iden-*

tidad)" and people who have "lost their identity (*sin identidad*)." When
I asked her what she meant, she said that those con identidad who
were proud of their history, culture, and identity; it meant that people
reflected on their past and were proud of their traditions. Segura believed that Afro-Colombians who sought to assimilate into mainstream
society did so because they considered their past a *"retraso* (backwards
step)." Such people were ashamed of being black and therefore were sin
identidad.

For Segura and many others associated with the PCN, mainstream
black politicians were sin identidad, despite their rhetoric of acting in
the interest of Afro-Colombians. But unlike Segura, many others from
the PCN labeled sin identidad all Afro-Colombians at the CVC meeting
who were against the PCN's position. This surprised me, since several
non-PCN Afro-Colombian activists were also celebrating black identity.
For instance, as the young rappers from Grupo Ashanty said to their
CVC audience, their performance was meant to celebrate elements of
black identity that resonate with urban black youth.[2] After their performance, the troupe invited everyone to join them in a "March against
Racism" organized by Cimarrón and UniValle students to be held in a
few weeks.

At the CVC meeting, many *caleños* (people from Cali) and PCN members from Buenaventura and the coast were attired similarly—in T-shirts
and hats depicting international heroes of black and nationalist struggles
(Nelson Mandela, Martin Luther King, Malcolm X, Mahatma Gandhi,
Bob Marley). Many of these young activists proudly sported dreadlocks
(*trensas*) even though they were not considered fashionable (in urban
areas) or "decent" (in rural zones). Both sides handed out flyers for black
events, articles about the U.S. Civil Rights Movement, and the South African antiapartheid struggle. In their speeches at the CVC meeting, PCN
members and non-PCN members drew on similar ideas of black history
and struggle even as they invoked different symbols of black culture.

Yet such similar views of culture did not make it any easier to form alliances or supersede differences to organize for black rights. At the CVC
assembly, PCN members opposed the candidacy of black politicians to
the CVC's board of directors because of their links to major party politics
and their acceptance of the "government's ways of doing things." PCN
members also argued—legitimately, I believe—that while riverine black
communities would be most affected by CVC's development undertakings, their concerns were unlikely to be represented by party politicians
hailing from Cali. Finally, they were convinced that their candidate,

Jorge Aramburo, was more committed to the struggle for black ethnic and territorial rights and to an alternative, ethno-cultural politics.
Aramburo, who had adopted the African moniker Naka Mandinga, was
from a mining community in Río Yurumanguí, near the Valle del Cauca
coast. PCN members argued that party politicians like Agustín Valencia
and his cronies were manipulating or buying the votes of students and
Aguablanca dwellers.

It may well have been true that the career politicians attempted to
buy caleño votes. But the PCN's vision of ethno-cultural politics, based
on perceptions of coastal Pacific realities, did not resonate with urban
blacks. During the fracas at the CVC meeting, many caleños argued that
Aramburo had neither the political experience nor a sufficient understanding of the issues faced by the increasing numbers of blacks living
in urban and peri-urban areas. By some estimates, by the 1990s about
45 percent of the population of the Pacific was concentrated in regional
centers such Buenaventura, Tumaco, and Quibdó (Urrea Giraldo et al.
2001). PCN leaders were aware that the realities and needs of this black
population, which included recent migrants from regional centers and
rural areas, were different from those faced by black communities living
in the Pacific littoral. However, they believed that rather than representing Afro-Colombian interests, existing mechanisms of politics and
economics in Colombia were largely responsible for the massive, violent
displacement of blacks from their ancestral homes in the riverine and
coastal regions of the Pacific.

The PCN saw in Law 70 a set of mechanisms favorable to its ethno-
cultural agenda. Article 47, Chapter VII, of Law 70 decrees that black
communities have the right to development according to their culture,
and under Article 57, a study commission was set up in October 1994 to
formulate a Development Plan for Black Communities (Plan de Desarrollo para las Comunidades Negras, PDCN). PCN activists saw the proposed PDCN as an opportunity to draft a grassroots-based and "culturally appropriate" model of development for black communities.

Culturally Appropriate Development:
Formulating the PDCN

Critical of prevailing development paradigms, PCN activists did not believe that the needs of black communities could be met within the parameters of Plan Pacífico and Proyecto BioPacífico—or, indeed, that

they would prove compatible (Organización de Comunidades Negras 1996; Proceso de Comunidades Negras n.d.; Rosero 1995, 1996). They characterized the latest plans to extract natural resources, harness "human capital" and insert the region more firmly into the market economy as extensions of past strategies (Grueso 1998; Grueso et al. 1998; Rosero 1996). They viewed the institutional and market dynamics of this development as harbingers of "conflicts over resources, violence for access to land, clientelist politics, fights over bricks and mortar, trading ancestral customs for consumerism, money for life options, and the deterioration and contamination of the environment" (Proceso de Comunidades Negras n.d.: 2).

Rosero also critiques the state's representation of the region and its people as "marginal and impoverished" and contends that state development plans are a form of "development without subjects" because they ignore the perceptions, actual demographics, needs, and agency of black communities (Rosero 1996: 182). In February 1995 at a meeting in Cauca, Rubén Hernández, a PCN member from the Atlantic Coast, noted that despite recognition of blacks as a distinct group in Law 70, nationalist, integrationist discourses promoting the formation of black "citizens" to participate in the emerging economy were still strong. Hernández argued that black communities needed to gain a better understanding of the larger political-economic context of the struggle for black ethnic rights to propose an alternative development model. For the PCN, the alternative to capitalist mega-development and its ills lay within local systems of economic production. They believed that a PDCN should be informed by, and rest on the perceptions and realities of, the grassroots black communities of the Pacific.

It was to formulate such an alternative strategy that black activists gathered in the small town of Buenos Aires, in the hills of Cauca, in February 1995. Among the seventy or so attendees were members of three regional *palenques*: Valle del Cauca (the OCN), Cauca (Cococauca), and Nariño (Palenque Regional del Nariño). Also present were several black activists and professionals from the Atlantic coast and Cali. Given post-Law 70 splits in the black movement, no Chocoan representatives attended. I traveled to the meeting with Yellen Aguilar and his mestiza compañera, the anthropologist Camila Moreno. Moreno worked with the Social Solidarity Network (Red de Solidaridad Social) President Samper's welfare program, which replaced the National Rehabilitation Plan. In keeping with its mandate of aiding the country's most economically marginalized populations, the Social Solidarity Network had recently launched an Afro-Colombian initiative. Moreno headed the regional office in Buenaventura. Although the palenque was an espacio propio, Moreno and I attended as participant observers.

Historically, palenques were fortified communities of escaped slaves.[3] During the trip to Buenos Aires, Moreno and Aguilar explained that, for the PCN, palenques symbolized autonomous spaces for cultural reconfirmation and an independent territory for physical expansion (PCN–OREWA 1995: 33). At Buenos Aires, Hernan Cortés, a historian and a key figure of the Palenque de Comunidades Negras in Nariño, explained why the PCN had chosen the term "palenque" to refer to its regional organizations:

> The organizations in the southwest departments have a local or regional flavor. History shows that the process of land recuperation is not new but has been going since the first palenque. Therefore, we call this the palenque—we see it as a political struggle. It is a space from which we administer the struggle for land rights. The palenque is not just an organization but a continuation of what our ancestors started. It takes the place of a traditional, ethnic authority.

The term "palenque" may refer to any and all of these: specific organizations, espacios propios, spaces of resistance, and spaces of organization and cultural revindication. We arrived in Buenos Aires on the afternoon of Friday, February 17, and joined a team working on the PDCN. Aguilar began the session by remarking with brutal honesty, "We [the PCN] have not constructed a concept, an idea, a plan of development. Despite our rhetoric we have not conceptualized what we want very clearly." He noted that there was chaos within the Subcomission on Development of the High-Level Consultative Commission and that the drafts of the PDCN circulating in official circles were produced by technical consultants and, not surprisingly, paralleled the state's development logic.

Alfredo Riascos from the Nariño palenque attributed the problems in the consultative commission to the absence of "institutional memory" and weak links among the various black movements, noting:

> There is little continuity between the special commission (of AT 55) and the High-Level Consultative Commission, although the former had also debated the issue of socioeconomic development among black communities. The current members of the Consultative Commission must meet with at least a few of the ex-members of the special commission.

Furthermore, the Consultative Commission does not know what is going on in the Division for Black Community Affairs, and vice versa. For example, division members are passing on to the state any PDCN drafts they receive without consultation or discussions with the Consultative Commission.

Rubén Hernández and Alfonso Cassiani from Cartagena and Marilyn Macado from Cali echoed Alfredo's observations. Rubén later suggested that the PCN needed to define and clarify its relations with the Division for Black Community Affairs. Alfredo ended his comments by lamenting the *Chocoanización* of the process to draft the PDCN. By Chocoanización he not only meant that groups from the Chocó were pushing their drafts of the PDCN, but also that these socioeconomic plans were formulated according to the clientist logic of Chocóan politics.

On the second day at Buenos Aires, Luis Mosquera, from a Cali-based black-rights organization called Kumahana, gave an impassioned testimonial against the intrusion of capitalist modernity in the Pacific, noting that:

> International interests and especially capital from superpowers such as North America and the Asian-Pacific Rim want to penetrate the Colombian Pacific. Opening markets and promoting economic growth is a palliative. There will not be space for alternative philosophies, religions, and beliefs. This is the contradiction of capitalist expansion in the region. The Western superpowers—the United States and Europe—want to protect natural resources and biodiversity because they know that these are the future sources of raw material for capitalist forces. So they promote culturally sensitive, sustainable development projects as a form of self-protection.

According to Mosquera, it was because of such dangers that even urban black communities had to be committed to a broader vision of the PDCN.

Gabino Hernández, a lawyer from Cartagena who was then based in Cali, spoke about the increasing presence of armed forces (the military, drug traffickers, paramilitary forces, guerrilla groups) in the rural Pacific. He asked his compañeros to reflect on how negotiations between the government and armed groups were likely to affect the PDCN. Attesting to Mosquera's concerns, Hernández reiterated the importance of understanding the articulation (*coyuntura*) of the economic and political forces in Colombia while attempting to formulate alternatives from the perspective of rural black Pacific communities.

At this point, someone reminded the group that its task was to generate some concrete proposals for a PDCN. Rosero urged the assembled company to remain committed to its strategy of developing the PDCN in conjunction with grassroots black communities. Remarking on the differences in the material needs and immediate realities of urban and rural communities, he also asserted the importance of conducting workshops with urban communities in cities such as Cali and Bogotá and regional centers such as Buenaventura, Guapi, and Tumaco. However, Rosero agreed with Cassiani that the PDCN was more than a blueprint for Afro-Colombian development: it was better understood as a cultural *and* political manifesto, an opportunity to reconceptualize the notion of *black communities* on the basis of shared history and culture and to reaffirm a *collective* black identity based on the material and symbolic connections to a particular black homeland or territory in the Pacific. That is, PCN activists seemed to agree that the institutional spaces opened up by Law 70 and the PDCN should be used by black communities to critique the state's plans and mega-development models and to elaborate alternative visions for their future and that of the Pacific. The working group ended with more questions than specific suggestions for a PDCN.

Several such working sessions were held at the Buenos Aires palenque, where discussions about "ethno-education," "environmentally sound resource management," "land titling," and so on went on late into the night. In the plenary session, I learned that most had ended as inconclusively as the PDCN debate. At the palenque's close, the task of bringing together these reflections into a concrete PDCN remained.

The PCN did, at least, take steps toward concretizing its own vision for a PDCN. A focal point was the subsistence practices of black communities living in the Pacific—agriculture, artisanal fishing and hunting, shellfish collecting, extraction of various renewable non-timber products, and small-scale logging and mining. PCN members noted that Afro-Colombians had lived in the region since colonial times and have developed these ecologically sustainable systems of economic production based on African and Amerindian traditions. For the activists, these "ancestral" practices and the associated social relations based on kinship and family ties were the locus of black ethnicity and difference and lay at the fulcrum of "culturally appropriate" development in the region.

Rivers were often considered the logistic and conceptual loci of such organization. Black households strung out along a given river in the Pacific often consider themselves part of a single community. Connections to a river are reflected in every aspect of the quotidian practices of black communities—people live on its banks; fish its waters; farm, log, hunt, and engage in other subsistence activities centered on the territory it defines. Or, as expressed in the Perico Negro document, resource appropriation and the Afro-Colombian notions of territoriality are based on "the natural ebbs and flows (of the rivers and the sea) and of the cultural flow of products and people (land use practices, migrations, barter/ exchange)" (PCN–OREWA 1995: 30). Since the river is often a primary focus of self-identification, each community or set of communities along a river was encouraged to organize community councils (*consejos comunitarios*). Law 70 also calls for the formation of such councils as local-level administrative units to manage collective black lands. However, the PCN saw the community councils as having a broader function. Starting from a positive sense of their ethnic identity, each council would be required to:

1. Discuss their own socioeconomic realities and how they would be affected by economic development in the region. After analyzing their situation they would articulate their own needs and generate proposals to meet them.
2. Generate plans to demarcate their lands and to manage the natural resources therein in ecologically sustainable and culturally appropriate ways.

The PCN hoped that such proposals would provide concrete expressions of ethno-cultural alternatives to state-sanctioned capitalist development.

Next, representatives from these local-level councils would gather in regional palenques (*palenques regionales*) to continue discussions and negotiate plans at the regional level. All regional palenques would then convene in a National Council of Palenques (Consejo Nacional de Palenques, similar to the one at Buenos Aires) to coordinate regional plans and formulate proposals for ethno-education, socioeconomic development, preservation of cultural identity, collective land rights and natural-resource management on black lands, and other issues outlined in Law 70. Various *equipos* (teams) would discuss and clarify the parameters of "ethno-education," "environmentally sound resource management," "land titling," and so on, and would provide logistical or operative assistance at each step.

PCN members repeatedly claimed that their organizational strategy was democratic and *necessary*. They argued that, in contrast to the "vertical" structure of institutional policymaking procedures, the various espacios propios would ensure that all members of the black communities could express themselves and engage in dialogue. Or, as Orlando Pantajo of Cococauca told me, "Black communities must convoke, convoke, convoke. After the fifth or the fiftieth time we will reach some agreement."

Only after reaching such collective agreements could conversations with state officials and NGOs involved with issues concerning black communities move forward. Specifically, *palenqueros* would represent their concerns in espacios mixtos (mixed institutional spaces) such as the High-Level Consultative Council and departmental councils (*consultivas departamentales*).

However, as numerous gatherings made clear, such a model of democratic participation did not always work in practice. Among the deficiencies were chronic shortages of time, funds, a capable cadre of activists dedicated to such an organizational strategy, and black and non-black entities willing to engage with such an approach. Then there was the diversity of cultural practices and land-use forms and the difficulties of organizing widely dispersed rural communities. I observed these challenges firsthand during field visits in rural communities in all four Pacific states.[4]

Life, Land Use, and Organizing in Calle Larga, Río Anchicayá

Between July and October of 1995, I spent several weeks in Calle Larga, a small village along the Anchicayá River in Valle del Cauca, a few hours from Buenaventura. In the following two sections I discuss the dynamics of el proceso based on my field work and my conversations with Natividad Urrutia, one of my key informants and one of several local supporters of el proceso.

Natividad—or Doña Naty, as she was universally called—lived in Calle Larga with her husband, Don Protacio (Don Poty), two of their ten children, and four grandchildren. Unlike many Calle Largeños, who engaged in multiple activities (farming, fishing, mining, logging), the Urrutias considered themselves primarily campesinos or *agricultores* (peasants or farmers). They raised a number of subsistence crops typical of the region: *papa china* (an araceae of Asian origin with a high starch content), corn, yucca, and sugarcane. They also planted a variety of fruit trees—bananas, *borojo*, cacao, mango, citrus, coconuts, *chontaduro* (peach palm)—in the *finca* (farm) and *monte rozada* (cleared forest) around the house. Behind the monte rozada was the *monte* (uncleared forest) from which the Urrutia family collected firewood and timber for household use and where the son-in-law occasionally hunted. Doña Naty's now deceased father had excavated a *zanja*, or canal, in the monte to transport the timber he logged and to gain access to his small and isolated gold mine. Other kin used the zanja with Doña Naty's permission; at times, she rented it to non-kin.

Like many other rural black women, Doña Naty had a vegetable garden and an azotea (raised garden) of culinary and medicinal herbs, reared a few chickens and pigs, and made guarapo (a fermented drink) and *biche* (a distilled liquor) from sugarcane to sell to loggers and miners passing along the river. Doña Naty added to the meager household income by selling soap, rice, oil, sugar, kerosene, and miscellaneous things from a small dry-goods shop in the house. She was also a community leader, midwife, storyteller, and *coplera*.[5] She told me that she had learned the art of the *copla* (as well as many songs and stories) from her father, a well-known griot. But she was also a composer in her own right. During the time I spent at her house, she often sang her coplas and *arrullos* while playing a *guasá* (a musical instrument made of bamboo) that she had made herself and showed me songs she had written down in a limp, dog-eared notebook. Among the latter was the following copla about life in Anchicayá, the black communities, and the black movement:

CUANDO VAMOS A LOS EVENTOS	WHEN WE ATTEND EVENTS
Cuando vamos a los eventos	When we attend events
Pensemos que vamos a hablar	We think we will speak
Porque somos de la costa	Because we are from the coast
Que sofrimos todo el mal	We suffer from ills
Somos los Anchicagueños	We are Anchicagueños
Que nos vamos a acabar	And we will disappear
Porque ya nuestra cultura	Because our culture
Se nos quiere terminar	Is what they want to destroy
Ya no tocamos marimba	We no longer play the *marimba*
Ya no bailamos la jota	We no longer dance the *jota*
Ya toda nuestra cultura	Now our culture
Se ha acabado en nuestra costa	Is disappearing from our coast

Soy costeña hasta la muerte
Pertenezco a Anchicayá
Es el río que se ha quedado
Sin poderse sustentar

I am a *costeña* until death
I belong to the Anchicayá
It's the river that has been left
Unable to sustain us

Los costumbre se acabaron
No las podemos usar
Porque se viene el invierno
Y acaba con la mitad

Customs disappear
We cannot use them
Because winter is coming
To break us in half

Quiero que los caporales
Vayan al río a mirar
Se acaba la papachina
También los jardinerales
Y quedamos otra vez
Sufriendo calamidades

I want the leaders
To go to the river to witness
That the papachica has vanished
Also the gardens
And still we remain
Suffering calamities

La gente de nuestra costa
Queremos complementar
Que la mina y el pescado
Lo pagan por la mitad

We the people of our coast
Want to compliment [our leaders]
For paying just half
What our mining and fishing's worth

Nos vamos a las montañas
A sufrir calamidad
Y ni si quiera nos preguntan
Cuánto vamos a cobrar

We go to the mountains
And suffer calamities
And they don't even ask us
The costs of our labor

Cuando venimos de allá
Y nis vamos a comprar
Ni tan siquiera se acuerdan
Que les dimos a ganar

When we return from there
And need to buy goods
We can't even recall
What they were to pay us

Nuestro hijos en las casas
Los mata la calamidad
Por que el salario que hacemos
Ya no nos quiere alcanzar

Our children at home
Are killed by diseases
Because the salary we make
Is no longer enough

Queremos seguir gritando
Para que nos oigan hablar
Que el negocio que traen
Los vamos a evaluar

We want to continue shouting aloud
So that they hear us speak
That the business they bring
We are going to evaluate

Cuando vamos a las tiendas	When we go to the store
Donde vamos a comprar	Where we buy our goods
Los precios están completos	The prices are too high
Y a nosotros no nos pueden rebajar	And they will not lower them for us.

The copla beautifully expresses Doña Naty's love for her riverine home and her pride in her culture. It also laments the fate of the river and the difficulties of the daily lives of the Anchicagueños, conveying her fears for her children's future. The copla further reflects on the bad consequences to come when the Anchicagueños become wage laborers. The subsistence lifestyle is hard, but being forced off the land to work for others is not a good option, either. Doña Naty's copla indicates that the benefits of development are dubious, at best.

One day while walking with Doña Naty, I had glimpses of the complex norms of resource use and land ownership that prevailed in Calle Larga. The gaggle of Urrutia children who accompanied us during our walk picked fruit from various trees in the monte rozada. Occasionally they skipped a tree, even though it had ripe fruit on it. When I asked Doña Naty about this, she explained that those particular trees and its fruit belonged to her sister-in-law downriver in the community of El Bracito, who would be coming up later to collect the fruit. Doña Naty in turn would be heading to El Bracito to harvest coconuts and tree snails from three hundred trees that she said belonged to her deceased father. During that time, the Urrutias would also harvest fruit from, and weed around, the eighteen coconut trees, sugarcane, and banana trees that they had planted in the El Bracito. I was a little taken aback at the idea of dead people owning trees. But Doña Naty assured me that the owner of the trees from which she had the right to collect fruit and mollusks was indeed her dead father. She said the ownership of the trees would pass on to her after *she* died, at which time the right to collect fruit and snails would be inherited by her brother. However, the trees that Doña Naty and her husband had planted would belong to their children (she did not say which ones).

While this exchange clarified who owned which trees and who had the right to collect fruit and snails and when, it was unclear who owned the plot of land in El Bracito. Conversations with other residents in Calle Larga, and three participative mapping exercises, revealed that all the families in the community owned or used land beyond Calle Larga. Conversely, several plots in Calle Larga belonged to people from other villages on the Anchicayá and neighboring rivers.

Such multifaceted forms of land and resource use are not unique to Calle Larga. Extensive anthropological and geographical research among black communities (del Valle 1995; del Valle and Restrepo 1996) reveals that property rights over forest resources such as trees, wildlife, and non-timber forest products, as well as fisheries and mineral deposits in the Pacific, are contingent and exceedingly complex. The PCN was surely right that dominant models of economic development and land tenure could not account for such local intricacies. But accounting for them within the PCN's own efforts—whether in organizing or in drafting a development plan for the black communities—was also an enormous challenge.

In July 1995, a team of researchers from the conservation organization Fundación Herencia Verde of Cali convened a meeting in Calle Larga. Fundación Herencia Verde had been working in the Anchicayá watershed for many years, and one of its latest projects was to help Anchicagueños draft a natural-resource management plan. On July 15, ten Calle Largeños met with six Fundación Herencia Verde members in the local school, a crumbling one-room brick building. After introductions, the Fundación Herencia Verde folks explained why they were on the river. Then Don Hector, the logging inspector from El Bracito, began speaking about logging problems in the area. The slightly inebriated Don Hector went on at length about how the PCN might be well versed with Law 70 but had little power vis-à-vis the state. He said that the prohibition on commercial logging after AT 55 was passed was hurting small-scale loggers, and despite the scarcity of good timber, the campesinos did not get fair prices for their wood.

Doña Naty corroborated Don Hector's concerns regarding fair timber prices and his observation that many people had lost confidence in the PCN because it did not help with such immediate concerns. But she defended the PCN, saying that it had a sound sense of what was going on in the Pacific and would know how to help the campesinos better if PCN members lived among them. But she and her neighbor Doña Esperanza also blamed campesinos for not keeping themselves informed, not participating in meetings, and not voicing their demands.

At this point, Don Miguel, a community leader quite as active as Doña Naty, spoke up in defense of the campesinos:

> The Fundación [Herencia Verde] works with me or Natividad or Esperanza or others. They bring the results of their work back to discuss them with us. We, the campesinos and Comunidades Negras [the PCN], are at fault if we do

not work with the people who come here to help us and work with us. People of the community understand and interpret or misinterpret things the way they want. Other people—Pablo and Kiran—have come with the Fundación to continue its good work. The Comunidades Negras has come and gone, but the Fundación is still here.

At this, I explained that I was with neither the Fundación Herencia Verde nor the PCN but was interested in understanding their work. I also explained the difference between the mandates of the Fundación Herencia Verde and the PCN and spoke about the latter's efforts to obtain broad ethnic and territorial rights for black communities. Don Hector accepted my points but brought up logging issues again. "Trees and timber won't disappear, but campesinos who are not allowed to earn their living by logging will." Jorge Ceballo, who managed Fundación Herencia Verde's agro-forestry projects, and Chucho, a schoolteacher and catechist from Calle Larga, both said that black communities needed to have a vision of collective land rights beyond the right to log. But, Chucho continued, "Collective land titles do not include forest land or rights over subsoil resources. So what does it mean for us to be owners of the land?" By this time, the meeting was coming to a close, and Doña Naty spoke again: "We need to be more clear, more animated about what we want to do. We have to know how to be owners of our land. We need to go to other places, to regional or national meetings to find out what is going at those levels and how it affects us. People tell us many things and make many promises but they do not keep them. We must make them keep their promises." Don Miguel suggested that the community convene again to continue these important discussions and figure out what to do next.

Two weeks later, a dozen people gathered at the school, including two representatives of the PCN. After the customary introductions, Doña Naty gave an overview of the ethnic rights guaranteed under Law 70. Chucho spoke about the constitutional limits of collective land titles and resource use. The rest of the meeting was spent discussing various norms, rules, and practices of resource use in the community and how traditional practices were disappearing. Don Hector spoke about logging again. Doña Naty spoke about producing and harvesting *borojo*, *naidi*, *chondaduro*, *papachina*, and other agricultural products. Don Miguel spoke about disappearing game and fish. He attributed the decrease in stocks to the use of dynamite in fishing, dust pollution from the commercial sawmill, and mercury from mining operations upriver.

Don Felipe, Doña Esperanza's husband, noted that miners like himself were also having a hard time making a living because they were not getting good prices for their gold. Other community members nodded in agreement and were keen to point out the problems caused by outsiders encroaching on their lands. But when Chucho said that they needed to determine how to put their knowledge into proposals and plans, the assembled company fell silent a while. Then most expressed that they felt confused about the terms of Law 70 and what it required of them. They specifically asked Fundación Herencia Verde to help them produce maps of their territory and draft a proposal for collective land titles, just as Armando of the PCN had requested. In the subsequent months, a Chilean Canadian master's degree student drew on "participatory rural appraisal" techniques to help Calle Largeños produce maps of their community. But the process of drafting a proposal for collective land titles based on these maps came to an abrupt halt a few weeks later.

In October 1995, several people were killed in the community of Sabaletas, en route to Calle Larga. Guerrillas and paramilitary groups began appearing regularly in the upper Anchicayá valley and making it difficult to travel along the river. For the rest of the time I was in Colombia, we received only sporadic reports from Calle Larga. Although unconfirmed, they were depressingly familiar: illegal logging and mining had accelerated; strangers were coming into the community to incite violence and steal land; many people were hungry; and many people were being forced to move away from the river.

Local Differences and the Cultural and Political Dilemmas of Grassroots Organizing

Although different in detail, Calle Larga's local concerns and challenges were comparable to those in other parts of Valle del Cauca and farther south, in Cauca and Nariño. There were many like Don Hector and Don Miguel who were primarily concerned with land rights and timber prices. There were others, like Doña Naty and Chucho, who recognized that el proceso and the broader black struggles emerging from Law 70 were part of resolving these pragmatic matters.

But even active community leaders like Doña Naty were stretched when called on to respond to the many demands of organizing. Participation *does* cost, and community leaders have only so much time for cultural and political projects (such as those of the PCN), state institutions, and research projects like mine. Gender roles further complicate participation. Although both Doña Naty and Don Poty participated in the black movement, Doña Naty was the more sought after of the two. Doña Naty knew that Don Poty resented this. But unlike many other husbands, he seldom said anything to her; nor did he try to curtail her activism. However, a combination of his displeasure and the many roles she played in the household (wife, mother, grandmother) and community (midwife, petty trader, coplera) led her gradually to decrease her involvement in the movement.

Although the terms of Law 70 and those of the PCN's participatory model were well intentioned and progressive, they were putting enormous pressure on local communities to engage in processes that they felt ill-equipped to deal with. Some considered these processes *ajeno* (foreign). After a lukewarm reception in Vereda Partadero, a small hamlet a short boat ride from Guapi, Dionisio Rodríguez of Cococauca admitted that mobilizing isolated black communities dispersed along the extensive Pacific watershed was a particular challenge for them. These communities were often suspicious of people hitherto unknown to the area and often considered PCN activists "outsiders." Most were unfamiliar with the new black movements and found the PCN's rhetoric of ethnic autonomy and territorial control too abstract. Many were unaware of Law 70 and found its terms equally confusing and alien. For example, while community councils were supposed to represent a local decision-making body, all the communities I visited established such councils (if at all) only after having heard of them in reference to Law 70. Most of these communities required assistance to understand the law and draft proposals to obtain land and resource rights, however limited those might be. In some cases, local elites tried to appropriate new or existing structures of authority to assert their power or to air their grievances against each other. Overall, local tensions, economic pressures, power differentials, and the dynamics of social roles structured community-level organizing. And all of these pressures deepened with the arrival of armed forces in the region.

The PCN's goal of organizing an ethno-political movement and drafting a PDCN as an alternative model of culturally appropriate development faced yet another complication: that of identifying the people it hoped to organize. The discussion of life in Calle Larga offers a glimpse into the complexity of resource use and land ownership among black communities. Early historical and ethnographic research among

Afro-Colombians (de Friedemann and Arocha 1986; West 1957; Whitten and de Friedemann 1974) reveals that black cultural and economic practices vary by occupation, geographic location and region, and the degree of "non-black" (Amerindian, Spanish colonial, *colono*) influence on local communities. Research conducted in the 1990s (including studies commissioned by the Colombian Institute of Anthropology and History under the aegis of Law 70) provides further evidence of the heterogeneity of Afro-Colombian cultural traditions and identity. Calle Largeños, for example, thought of themselves as "campesinos" and used the expression "Communidades Negras" only to refer to the PCN. Eduardo Restrepo (1996) notes that blacks in the southwestern Pacific communities call themselves "*libres* (free people)" in relation to a series of other groups, such as *indios bravos* (indians in general), *cholos* (neighboring Emberá indians), *paisas* (Antioqueño traders), *serranos* (whites or mestizos from the Andean parts of Nariño or Cauca), and *culimochos* (whites or mestizos living on the coast). He notes that the term "tuquero" refers to black loggers in the southwestern Pacific region, whereas black loggers in the north are *madereros*. In the northern coast of the Chocó, blacks may call themselves *negros* in relation to *cholos* or other indians. In short, cultural self-identifiers are not universal but context-dependent and variable.[6]

Nor does the term "Afro-Colombian"—which came into common usage only in the 1990s—serve as a common identifier at the local level. Unlike urban-based Cimarróns, among rural blacks the reference to a common "African" ancestry often conjures up images of a return to a dark, unknown continent with which they feel no connection. In the 1990s, the PCN began to use "black communities" as a broad term, able to encompass diverse black identities and experiences. This was also a means of recuperating the term "black (*negro*)" and infusing it with a positive sense to oppose its prior racist connotations.

Questions about the boundaries of the "black communities" and just who the Afro-Colombian movement sought to represent were unavoidable. Ethnic difference and cultural rights continued to be understood in terms of indigenous experiences, since until 1991 only indians were recognized as a distinct group with special rights. Ambiguity about Afro-Colombians as an ethnic group raised questions about what constitutes "ancestral" rights, "traditional" practices, and "authentic" cultural identity and led to bitter debates—among scholars, activists, and state professionals, including many supporters of the movement—about "black ethno-cultural" models of development. Extrapolating a single model of Afro-Colombian development from such heterogeneity was all but impossible in cultural terms, as the anthropological literature of the past decade shows. But the failure to form effective coalitions across their internal differences and with other social struggles was also a fundamental problem of the PCN's political strategy.

The PCN knew that effective struggles against state hegemony depended on building alliances across heterogeneous black interest groups and with those sympathetic to the black movement. As indicated in chapter 1, it was just such alliances with the church, *sindicatos* (unions), NGOs, and state entities that led to the inclusion of AT 55 in the 1991 Constitution and the passage of Law 70. Many of these connections evolved in various ways both favoring and moving away from black struggles. But as the CVC incident showed, negotiating regional, class, and ideological differences was not easy especially in the short term.

Individual PCN members networked strategically and broadly. Libia Grueso and Carlos Rosero were active framers of the PCN's ethnocultural vision and among its most vocal spokespeople outside Colombia. Through them and others, the PCN made particularly strong international connections with intellectuals, researchers, policymakers, and activists of various stripes (anti-globalization, environmental, supporters of indigenous politics, etc.) to gain support for an "ethno-cultural alternative" to the state's development plans for the region. PCN members traveled widely in the United States, Europe, and Latin America, participating in academic conferences and engaging with antiracist and anti-globalization activists.

However, the story within Colombia was somewhat different. A few PCN members (Aguilar and Grueso among them) worked for the state and NGOs. Since the early 1990s, Grueso had held jobs with state-run projects (such as Proyecto BioPacífico) and a variety of state and private agencies (the Ministry of Education, the national parks authority, FES, the World Wildlife Fund). In our many conversations, Grueso expressed that she saw her work in these organizations as crucially important, not only because it helped support her family, but also because it gave her space to do the political work of the black movement from within these institutions. Many other professionals within the PCN also held long- or short-term contracts with various state entities and NGOs. As an organization active in black struggles, the PCN, too, was necessarily involved with these entities and their myriad ventures in the Pacific. Yet according to Grueso, rank-and-file PCN members perceived any links with institutions as taking time and political energy away from ethno-cultural struggles.

The PCN's relations with NGOs, even those with long-term commitments to the Pacific (such as Fundación Herencia Verde and Fundación Habla/Scribe) were tenuous. In September 1995, during a tense conversation with two Fundación Herencia Verde personnel about a participatory agro-forestry project on the Anchicayá River, Rosero explained that the PCN feared the "NGO-ization" of social and environmental struggles. For the PCN, the black ethno-cultural movement was primarily a political one. Technical and development approaches to implementing black rights were greeted with suspicion.

Consequently there was a de facto sense among PCN activists that el proceso was strictly closed to "outsiders." Not surprisingly, however, the boundaries between "insiders" and "outsiders," "blacks" and "nonblacks," those "with identity" and those "without identity" were far from clear-cut. Bitter divisions occasionally erupted in the PCN about the role of people like Camila Moreno, Aguilar's mestiza girlfriend. Moreno and many others who worked in the Pacific were strong but critical allies of black struggles. The PCN maintained a rhetorical autonomy and distance from many of its supporters (including, as noted in my methodological remarks, myself), even as individual PCN members engaged with them and the movement relied on their solidarity.

The PCN's ties to groups or people, including black groups, who did not fully subscribe to its organizational strategy were also ambiguous. Youth groups such as Juventud 500 in Buenaventura, cultural groups such as Ecos del Pacífico in Tumaco, community groups such as Fundación Atarraya in Guapi, and numerous women's groups in all four states were among the local organizations that gained new momentum in the 1990s from the rhetoric of black "ethnicity." Many of these groups remained at least nominally linked to the PCN. They agreed with the PCN's principles and supported its goals of building a collective black movement. But they also had their own specific aims and mandates, and their organizational strategies often differed from the PCN's. Working in conjunction with the state, NGOs, or other entities, they developed ideas of black identity and cultural politics in interesting and creative ways. I examine some of these developments in chapter 4.

In the meantime, state initiatives such as Plan Pacífico and Proyecto BioPacífico unfolded with greater rapidity, even as the PCN denounced them. Chocoan groups, black politicians, and new black factions increasingly engaged such initiatives. Claiming to represent black communities, these *politiqueros* also formulated their own version of the PDCN (Klinger 1996). William Villa (1995) and Rosero (1996) remark bitterly that this PDCN, replete with economic statistics and social-welfare figures, parallels institutional "development-speak" and posits the complete integration and insertion of black communities into the modernization schemes of the state and capital.

The appropriation of the PDCN process by Afro-Colombians aligned with mainstream political and development sectors raises a fundamental question about the PCN's ethno-cultural strategy: how to reconcile its insistence on a black-ethnicity-specific but still undefined position with the imperative to act *within* the existing political processes.

Politics of the PCN: Contradictions and Consequences

At his talk in UniValle in August 1995, Rosero reflected on the PCN's ideological vision and organizational strategy. He stated that the PCN's vision emerged from reflections on the leftist discourse of Cimarrón and the demands for "ethnic revindication" during the constitutional-reform process. According to Rosero, some twenty to twenty-five organizations—women's, youth, and cultural groups—took Cimarrón's discourse of racial discrimination, economic marginalization, and human rights as their point of departure, even though they were organized around their own issues. Many PCN members, including Rosero, had been part of Cimarrón and its precursor, a student-led radical study group (Wade 1995). Although the PCN asserted its distance from Cimarrón, its analysis of how the state and forces of global capitalism shape the regional political economy were influenced by the left. Rosero, Grueso, and other PCN intellectuals were also well versed with—and, indeed, involved in—leading debates about culture, politics, and economy in Latin America. This familiarity with intellectual discussions, combined with an engagement in social struggles, informed their critique of the inadequacy of marxist analyses in accounting for the cultural specificity of economic marginalization.

Like many other social movements, black movements in the 1970s and 1980s were inspired by ideas of revolution and struggle for economic and political justice. From the 1980s on, attention to cultural rights began frequently to appear in considerations of development and democracy in Latin America and in the claims and discourses of black movements.

Many Afro-Colombian activists drew on the tenets of the new constitution to make their case for black rights and equality, but according to Rosero (1995), "organizations that talk about black marginality and needs for equality are not focusing on the problem of the specificity and heterogeneity of black communities. Nor do they talk about redefining social relations between blacks and the rest of society. There is no talk about autonomy, only calls for social and economic equality, without examining the nature and contents of this equality." However, Rosero admits that "equality" and "the right to be different" are two sides of the same coin. Noting that a political strategy from either extreme was risky, Rosero argues for a collective black movement containing within it a diversity of identities and positions.

The historian Joan Scott's discussion of the "equality-versus-difference" debates in feminist politics is useful here (Scott 1990). Scott notes that, when paired dichotomously, equality and difference present an impossible choice. She claims that feminists cannot give up "difference" because it is one of their most important analytical tools. Nor can they (or Afro-Colombians) disclaim the struggle for equal rights—at least, not as long as they function within current liberal, democratic political systems. Scott's suggested response involves two interrelated moves: the unmasking of power relations that pose equality as antithetical to difference, and the rejection of the "consequent dichotomous construction of political choices."

The former move requires us to take another look at how equality and difference are considered within liberal political theory and its language of universal justice and human rights. The political theorist Michael Walzer notes that "the root meaning of equality is negative; egalitarianism in its origins is an abolitionist politics. It aims at eliminating not all differences, but a particular set of differences, and a different set in different times and places" (quoted in Scott 1990). Within such an understanding of equality, argues Scott, different people are considered equivalent, not identical, for the particular purposes of democratic citizenship. Thus, at different times such categories as independence, property ownership, race, age, and sex have served as measures of equivalence. For Scott, this suggests that the political notion of equality not only includes but depends on the existence of differences.

Scott's analysis breaks free of the binary coupling of equality-versus-equivalent difference and makes room to ask why and how the meanings of difference are constructed. It is this particular questioning of the meaning of identity that Rosero invokes when he suggests that equality and difference are two sides of the same coin.

New theories of social movements also underscore how the construction of collective identities structures the new "politics of resistance" such as the PCN's ethno-cultural politics. Both the 1991 Constitution (explicitly naming Colombia a "multiethnic and pluricultural" nation) and Law 70 have provided opportunities for the black movement to reformulate the meaning of black identity and redefine their relations with the state. As documented in this book and by the activists themselves (Cortés 1994; Grueso et al. 1998; Hurtado 1996; Organización de Comunidades Negras 1996; Rosero 1995), the PCN's post-Law 70 efforts to rally around "the right to be different" and to live autonomously in the Pacific as blacks were also a strategic imperative: a conscious resistance of neoliberal globalization and apertura.

Diana Fuss (1989) and Gayatri Spivak (1990), among other feminist theorists, have suggested that at certain times, self-critical claims to a strong, shared identity may be useful as loci of political struggle. It may be necessary, they argue, to "risk essentialism" by invoking a repressed, undervalued, or de-legitimized category of identity within a dominant discourse—precisely so that it can be foregrounded. With respect to the PCN's ethno-cultural politics, Arturo Escobar (1997: 219) argues that the "essentializing" of black identity and the defense of riverine livelihood practices "is a strategic question to the extent that they are seen as embodying resistance to capitalism and modernity."

What are the risks in question? Among them is certainly a potential for reifying "strategically constructed identities" and romanticizing resistance. Escobar also cautions against this. "Although [black activists'] arguments are often couched in a culturalist language," Escobar writes, "they are aware that the intransigent defense of culture is less desirable than a cautious opening to the future, including a critical engagement with modernity" (Escobar 1997: 219). In his August 1995 talk at Uni-Valle, Rosero also seemed aware of the dangers of essentialism, particularly given the diversity of circumstances, values, and aims within black communities:

> Understanding the differences of black communities from other sectors, the differences within black communities, and coming to terms with the heterogeneity of interests depends on figuring out the specificities of black communities and articulating a position based on them. But this will take time and effort. We do not know how much time. . . . There is not enough reflection or

saved capital to respond to questions that are just coming up. This is causing a lot of confusion, what state officials call "noise," inside the black organizational process.

The "noise" inside black organizations is also a reflection of the complex ways in which social movements, the state, and development forces are intertwined. As the dilemmas with the PCN's attempts to draft a development plan for black communities illustrate, opposition to state modernization and globalization cannot be reduced to simple reversals: "local communities" instead of wider networks, "local traditions" instead of knowledge and practices gleaned within and without.

Neither can the PCN's refusal to accommodate the power of the state and economic forces be understood as the assertion of autonomous resistance. The PCN and black movements are firmly embedded within Colombia's attempts to be a "modern" nation—one capable of economic growth and allowing (indeed, seeking) an expanded participatory democracy (Agudelo 2001: 8).

Carlos Agudelo (2001), Mauricio Pardo and Manuela Álvarez (2001), and Peter Wade (1995) stress the role of structural factors, especially state policies and the new laws, in shaping the dynamics of black ethnocultural politics. Wade (1995) argues that not just state funds but also changing state forms, new laws on cultural rights, international treaties and policies which foster the use of "identity politics," structure the PCN's discourse and strategy. The latest development discourses (with their emphasis on environmental sustainability, human rights, and participation) also inform the strategic cultural politics of the PCN. Stefan Khittel (2001) argues that the state's "face of hegemony" is blurred because black activists who draw on and construct black history are also part of the state.

In demanding autonomy *from* the state, the PCN recognizes the state's authority and in a curious way legitimizes it by engaging its instruments (laws, policies) and institutions. Perhaps it is the recognition of this conundrum that leads PCN activists to guard espacio propios so jealously and to attempt to keep el proceso closed to outsiders. That it cannot not do so (in Spivak's phrase) is one of the inescapable aporias of development.

Another is that while the PCN critiques the state's agendas in the region, the state's interventions play a key role in shaping the dynamics of the black movement. After Law 70, black groups become increasingly dependent on state funds to support their mobilizations. Even as it denounced the government's intentions, the scramble for funds, and the corruption (or "Chocoanization") of the black movement, the PCN found itself at the mercy of those funds. Formulating plans for culturally appropriate development, forming community councils, and drafting applications for collective land titles and natural-resource management are also activities that blur the (already less than distinct) boundaries of espacios propios.

Cultural resistance and alternative developments are not simple possibilities. No local movement, including the PCN, can single-handedly halt the juggernaut of state and late-capitalist processes. The tripartite of ethnic rights, sustainable development, and biodiversity conservation served as an entry point for state interventions and expansion of bureaucratic power (Ferguson 1994). The push for economic growth, modernization, and development was a gateway for neoliberal globalization. But these same conjunctures also opened a window for ethnocultural politics and a call for alternatives to such modernity. "Local" movements around the world, like the PCN in the Pacific lowlands, attempt to disrupt the expansion of the power of ruling groups and capitalist forces and interrupt the depoliticizing functions of the "development" apparatus. These movements are structured as much by state and global economic forces as by local conditions. These dynamics are uneven and perhaps even overdetermined, but they cannot be reduced to structural effects.

❄ 4

"SEEING WITH THE EYES

OF BLACK WOMEN"

Gender, Ethnicity, and Development

From the window of the office of CoopMujeres, a women's cooperative in the southwest Pacific town of Guapi, I had a clear view of the river that gave the town its name. The scenes below were reminiscent of the ones I had seen two months ago in Quibdó: tuqueros punting long rafts of logs (hand-hewn with axes and lashed together with vines) to sell downriver; campesinos unloading bananas, coconuts, citrus, *borojo*, *chontaduro*, and other tropical fruit as well as fish and mollusks from their canoes; children bathing noisily on the riverbank while women washed clothes and pots nearby. A block away in the town's main plaza, the produce and fish market was in full swing. Lining the plaza were dry goods stores stocked with the basic items necessary for life in the rural, riverine areas of the Pacific: rice, sugar, rubber boots, twine, fish hooks, gasoline, rum. A stall selling local handicrafts, an initiative of CoopMujeres, was a recent addition.

I turned my attention to the bright, whitewashed room, alive with the energy of CoopMujeres members: several street vendors with whom I had haggled earlier that morning; Dora Ortiz, the *artisana* who had come down the stairs of the handicrafts stall so nimbly that I did not know she was blind; and Sylveria Rodríguez, the director of CoopMujeres, with her serious eyes and brilliant smile. There were two other visitors, the coordinators of a new Canadian–Colombian Program for Black Women. Sylveria was telling them about the future plans of the cooperative. Members wanted more training and help with income generating productive activities. In addition, they wanted workshops on the meaning of black identity and citizenship, political participation and leadership, women's health, and interpersonal and family relations.

FIELD NOTES, APRIL 3, 1995

Black women have always been active and visible in Pacific life,[1] but with the profound political changes at the end of the twentieth century, their activism and visibility took new forms. Besides being crucially important parts of ethnic movements, Afro-Colombian women, as the vignette from my field notes illustrates, were active members of development cooperatives. Emerging in the 1980s, these cooperatives became renewed targets of development attention in the 1990s. The juxtaposition of economic development and black ethnic organizing created new possibilities for *afrocolombianas'* (Afro-Colombian women's) activism. In the post-Law 70 period, Afro-Colombian women established autonomous networks of black women of the Pacific to address their gender needs and ethnic identity. While several PCN women activists were part of these women's networks, the PCN urged afrocolombianas to join with broader ethnic movements rather than organize independently.

In this chapter, I map how afrocolombianas drew on prevailing discourses of development and ethnic identity to assert their right to organize independently from state-led projects and from broader black movements while simultaneously being constituted as subjects of these forces. I argue that the networks of black women and black women's activism are neither autonomous expressions of alternative (feminist, ethnic, or non-Western) rationality for Pacific development and politics nor "instrument effects" of modernizing state power. Rather, by tracing the changing dynamics of afrocolombianas' organizing, I show how black women's subjectivity and agency are shaped differentially, unequally, and discursively by and against the political and cultural struggles.

Development and Black Women's Cooperatives

Like women in other parts of the world, black women in the Pacific organize and work collectively around their quotidian tasks. Many of these traditional forms of organization and social relations began to change with the waves of modernization in the Chocó (Escobar 1995; Lozano 1996; Rojas 1996). Within the first post-World War Two decade of Third World development, women, especially those in rural areas, were considered part of the economically "nonproductive" domestic and subsistence sectors. Within policy circles, they appeared, if at all, as welfare recipients or targets of population-control and poverty-reduction programs (Braidotti et al. 1994). In the early 1980s, these representations gave way to "Growth with Equity" programs, which focused on "harnessing women's labor" for economic growth and integrating women into mainstream development processes. Such programs were strongly influenced by Ester Boserup's book *Women's Role in Economic Development*, a landmark work published in 1970, which demonstrated that Third World women make a considerable contribution to productive sectors, especially in agriculture. These programs were framed by the Women-in-Development (WID) approach, which aimed to bring women up to par with men economically, and to ensure that they received equal benefits from development.[2]

These general trends were reflected in development plans for the Pacific, with women appearing as beneficiaries in the population, nutrition, and rural health projects of President Betancur's Integrated Development Plan for the Pacific Coast (PLADEICOP) initiative and of the Plan for Welfare Homes (Plan de Hogares de Bienestar), which was part of President Virgilio Barco's Social Economic Plan for the Pacific Region (Escobar 1995; Lozano 1996). While women were not the focus of agriculture intensification projects, Betty Ruth Lozano (1996) and Jeannette Rojas (1996) note that large numbers of women became integrated into the agro-industrial sector as low-paid menial workers on shrimp farms and African oil palm plantations. In 1988, probably influenced by WID approaches, an extended PLADEICOP launched a women's component to facilitate women's contribution to the productive sector (Lozano 1996; Rojas 1996).

In the 1990s, PLADEICOP received funds from UNICEF and other international donors to encourage rural groups to establish savings-and-loan cooperatives. The main aims of these cooperatives were to form large consolidated groups to qualify for loans, get technical help to improve artisanal and agricultural production, and obtain institutional support for marketing their products: in short, to increase rural incomes. Several cooperatives, including women's cooperatives, were formed with PLADEICOP's support. Among them were CoopMujeres (short for Cooperativa de Ahorro y Crédito de Mujeres Productivas de Guapi, or Savings and Loan Cooperative of Women Producers of Guapi) and Ser Mujer (short for Ser Mujer Cooperativa de Ahorro y Crédito, or Womanhood Savings and Loan Cooperative) in Tumaco.[3]

Ser Mujer came into existence when an association of eighteen groups of "bakers, seamstresses, community mothers, vendors of fish, shellfish and street food, charcoal makers, and handicrafts makers" came together in 1992. This association, the Tumaco Women Delegates Group (Junta de Delegados de Mujeres de Tumaco; JUNDEMUT), became a formal cooperative in 1993.

Up the coast in Buenaventura, the Foundation for the Development of the Women of Buenaventura (Fundación para el Desarrollo de la Mujer de Buenaventura), known as Fundemujer, emerged in 1989. Women's health, especially pre- and postnatal care for young mothers, was a central concern of Fundemujer. But the cooperative also supported women's productive activities.

Although most of the members were Afro-Colombian (given the demographics of the region), membership was not restricted to black women. Cooperative women trained in micro-enterprise management, established rotating, low-interest credit schemes, and formal social solidarity networks, including emergency funds to aid women in times of acute domestic crisis. Besides the usual bureaucrats and development experts, a few feminists and social activists such as Rojas and Lozano worked closely with these women's groups, often serving as consultants (*asesoras*) to local or federal state programs and regional NGOs. Rojas (1996) notes that in keeping with the broader development aims in the region, the work of cooperatives was directed toward addressing basic, material needs, or what she and others call the *tener* part of women's lives. Given the aims and structures of the cooperatives, however, there was little room for its members to discuss issues such as sexism, domestic violence, or ethnic and racial discrimination—regular aspects of experience though they were. Lozano (1996) considers these omissions a form of neutrality with respect to gender concerns within the cooperatives.

By the time I visited in 1995, women's cooperatives in the Pacific had an established organizational history behind them and seemed successfully institutionalized. CoopMujeres claimed 122 members, and Ser Mujer claimed 220. With twenty-five women's groups and a total of eight hundred members, Fundemujer was by far the largest cooperative. Income generation through productive activities was the central aim of most cooperatives, but with changes in the region's political economy and ethno-cultural dynamics, new concerns began to appear on their agendas. While not quite as close to home, the buildup to the June 1995 United Nations Women's Conference in Beijing and international discourses about gender, development, and the environment also had some bearing on the changes occurring in the cooperatives.

An important vehicle of change was the Program for Black Women, a new initiative funded through the Canadian and Colombian governments and Foundation for Higher Education (Fundación para la Educación Superior; FES), a Cali-based development NGO. The coordinators of this program were Rojas and Ana Isabel Arenas, the head of FES's Social Division, who had professional experience working with gender and women's issues. In 1994, the program began working with women's groups in Bahía Solano, Chocó, and funded projects on health, self-esteem, gender, and ethno-cultural identity, sustainable natural-resource use and management, and support for income-generating activities. The program's approach was implicitly informed by the Gender and Development (GAD) frameworks, which were replacing the integrationist WID perspectives and aimed to "empower" women to become key decision makers in the household and community. Framed with critical insights from feminists and development professionals in the north and south, GAD approaches aimed to transform existing gender relations by addressing women's practical (everyday, immediate—*tener*) needs and their strategic (long term, political—*ser*) gender interests and identities (Kabeer 1994).[4]

On April 3, 1995, I accompanied Rojas and Arenas on their visit to CoopMujeres to brainstorm ideas for potential projects under the Program for Black Women. After introductions, Arenas gave an overview of FES and its various social and gender projects. Then Sylveria Rodríguez, the director of CoopMujeres, spoke about the cooperative—its membership, its functions, and its future needs. Many *socias* (partners or cooperative members) wanted technical training to learn about obtaining credits, keeping accounts, and increasing the production and sale of their goods. Others chimed in with suggestions for workshops on women's health, family relations, political participation and citizenship, and so on. One member asked if the cooperative could sponsor a talk about women and the meaning of Afro-Colombian ethnic identity.

Joining the discussion, Arenas observed how gender often seemed to be a secondary element in conversations about development and floated the idea of self-esteem or women's rights workshops. Rodríguez responded, "we already *know* our rights. Now we need to learn how to *obtain* our rights, teach other people about our rights. We need to educate our men about women's rights. Last year we celebrated Father's Day, and many of our *compañeros* came. This year we are trying to make each member "win" her partner's support and bring him to the workshop." When CoopMujeres celebrated Father's Day, its members treated their attendant *compañeros* (in this context, the term means husbands or domestic partners) to a delicious meal of dishes made with regional products. The event showcased how cooperative members help sustain and recover local produce and culinary traditions. After the meal, there were humorous skits and songs about the various kinds of relations between men and women. One contrasted a home where sexism, jealousy, and abusive behavior abounded with another where the man helped with chores and child care and supported the woman's involvement in the cooperative. Many members agreed that their compañeros accorded them more respect when they brought more income into the household. The implication was clear: for members of the cooperative, more egalitarian relations with their compañeros were tied to economic security and income generation.

After an animated exchange about the achievements and aspirations of the cooperative, a consensus was reached regarding its next goals: to concentrate on income-generating activities, but also to expand the collective aims of the group to explore members' identities and rights as Afro-Colombians and women. Rojas and Arenas endorsed these ideas. They also discussed how concrete proposals, clear accounting, and accountability between partner institutions were crucial for the cooperative to achieve its aims.

On the following day, I met with cooperative members again. Among them was Cipriana Diuza Montaño, a schoolteacher who had recently joined CoopMujeres because it was "organized, grounded and actively helped single women, single mothers, poor women, and heads of families." But she confessed that after hearing about Law 70 and black rights, she was curious about "the fuss about black women." According to Diuza Montaño, the folks from Cococauca, the PCN affiliate in Guapi, who had organized the Law 70 workshop had "not told any stories about black women." When asked about her own understanding about being a black woman, she said, "We are black women, joyous but still enslaved, still afraid. We still need to learn to value our dialects, our religion, our dances."

Diuza Montaño's compañeras (in this context, the term means comrades, sisters in struggle, friends), including those from cooperatives in other Pacific coastal states, expressed similar sentiments. Despite their curiosity about black identity and cultural rights, these women were not involved with black regional organizations such as Cococauca or the

16

6 *Chapter Four*

PCN. This lack of regional political affiliation also characterized many members of Ser Mujer and Fundemujer. As women's groups, they primarily helped their members address immediate material needs. Some Ser Mujer members knew about Law 70 through the Nariño Palenque, but none were linked to the PCN or participated in el proceso. Overall, the group felt that the palenque's work was largely confined to rural riverine communities and had little to do with realities in Tumaco. In Buenaventura, a few Fundemujer members such as Mercedes Segura actively engaged ethno-cultural struggles, but as a group it remained autonomous from the PCN.

Despite the lack of close connections with black movements, the mobilizations around AT 55 and Law 70 created space for black women to link the issue of ethnic and territorial rights with their other concerns. Two regionwide black women's *encuentros* (meetings)—in Buenaventura in 1990 and in Guapi in 1992—had been organized by Fundemujer and other women's groups. Once again, feminists like Rojas played instrumental roles in organizing and obtaining funds through PLADEICOP, SWISSAID, and, later, the Program for Black Women. While the new Plan Pacífico also had resources slated for ethnic, gender, and environmental projects, these projects were still in the draft stage, and funds were not yet forthcoming. At the Guapi encuentro, black women expressed the desire to "create autonomous women's organizations that manifest and reflect our development, interests and ethnic cultural identity." The Black Women's Network (Red de Mujeres Negras) emerged there with aims "to establish communication and solidarity between different women's and mixed organizations, to promote women's organizations through education and empowerment, to strengthen ethnic identity, to study the realities and the needs of women and to make women aware of the management and sustainable use of natural resources and the environment." Participants in the network agreed to abide by the principles of "autonomy, affirmation of ethnic-cultural identity, respect, responsibility and conscience" (Rojas 1996: 218). Coop-Mujeres, Ser Mujer, Fundemujer and other cooperatives soon joined the network. In its first few years, the Black Women's Network established regional and local offshoots in Guapi, Buenaventura, and Bahía Solano. Women from black regional organizations such as the PCN also claimed membership in the network. More an idea than an institutional structure, the network is given shape when member organizations conduct activities in its name. Consequently, there are multiple understandings of its origins and purpose.

"Seeing through the Eyes of Black Women":
Engendering Ethnic Struggles

On August 23, 1995, I met with Patricia Moreno, Dora Alonso, Mercedes Segura, and Myrna Rosa Rodríguez of Fundemujer. Among other issues, we spoke about the relations between women's groups and ethnocultural struggles in the region. Recalling the experiences of the 1990 encuentro, the compañeras noted that the meeting had provided a space for black women to reflect on how they perceived their ethnic and gender identity, ask questions about their territorial rights, and discuss strategies to reclaim these rights.

After a brief pause, Segura continued, "And then we had to discuss how to work with our first enemy, the man-friend." Elaborating on her enigmatic statement, Segura said that the "first enemy" alluded not only to the menfolk at home but also to those within broader black struggles. Other Fundemujer members added that the cooperative helped in the Law 70 process, but that this was not recognized by the PCN. "We want to see ethno-political struggles with the eyes of black women," one member told me. "But the PCN does not want the two struggles together. Their position is that the gender struggle weakens the ethnic struggle."[5]

Naming Margarita Moya, Zulia Mena, Piedad Córdoba, María Angulo, Leyla Andrea Arroyo, Libia Grueso, and others, the cooperative members proudly reflected on the visible role of black women as leaders in current black struggles, including in traditional partisan politics. What was less visible, they claimed, was the support provided by black women and organizations such as Fundemujer during the mobilization process to ratify AT 55 and pass Law 70 and the support they continued to provide to black movements.[6] I, too, had noticed that it was mostly women who answered telephones, managed the books, and took care of running the offices at the PCN, Cococauca, and OBAPO. Many women in the PCN told me that it was mostly women who organized workshops, meetings and mobilizations, often on shoestring budgets.[7] They bought supplies, cooked and served food, and cleared up after gatherings. When there was no running water at the Bueno Aires Palenque, it was the women who figured out how to address the problem. Fundemujer compañeras and others said that women's logistical and administrative contributions tended to remain invisible or undervalued within el proceso.[8] Even an important PCN woman leader complained, in private, that male PCN members undervalued women activists' political and economic contributions.[9] Giving an example of a black activist who was once closely allied to el proceso, she also noted that women leaders who spoke too openly about women's concerns and gender inequality within the organization risked marginalization.

Although proud of black women leaders, activists outside the PCN felt that their visibility obscured differences within the black communities and unequal power relations between Afro-Colombian women and men. On the one hand, women were lauded for fostering a distinct black identity through their many quotidian tasks, such as growing and preparing specific foods, performing particular rituals, engaging in special health-care and healing practices. On the other hand, their essential positions in maintaining black family and community life served to justify or obscure their more "domestic" roles in current ethnic struggles.

Both women and men activists acknowledged that women's gender roles restricted their ability to participate in the more "public" or "political" aspects of ethnic struggles. They also conceded that women have distinct experiences and concerns—such as being primarily responsible for meeting the immediate material, or *tener*, needs of their families—arising from cultural understandings of their roles. But the movement found it difficult to address such "women's" needs. In Quibdó, Victoria Torres, a former executive committee (*directiva*) member of ACIA, told me, "The directiva was always too busy with other important issues, and the problems of women always got shunted."

The women Torres referred to were recent arrivals in Quibdó from the Atrato and other riverine communities. Often displaced by the very economic and structural forces the black movements fought, they struggled daily to find work and provide food and shelter for themselves and their children. These poor women seemed to fall through the gaps in both the state Social Solidarity Network (RSS) and the yet-to-be determined cultural-development alternatives of the black movements. It was to help such poor women that a few women in ACIA organized a Women's Section (Sección Mujer), which ran a small community kitchen (*cocina popular*) in Quibdó to provide some work and food for poor women. Women from ACIA and OBAPO also teamed up to organize a mass *invasión* (squatting) on the outskirts of Quibdó and to establish a neighborhood for displaced and destitute women. Torres told me that the Women's Section did not split formally from ACIA because it is committed to the black movement's ethno-political project and to implementing Law 70.

Torres and others from ACIA's Women's Section were also associated
with the Chocoan Women's Network (Red de Mujeres Chocoanas),
which had sponsored numerous events in March 1995 to celebrate International Women's Day in Quibdó. I participated in some of these
and spoke to Adriana Parra and Nimia Teresa Cuesta, both members of
the Chocoan Women's Network, who had organized a workshop titled
"Women: Equality and Development" on March 7. As part of the workshop, Parra and Cuesta asked participants to reflect on their notions of
development for black women. One participant observed, "The term
'development' brings visions of progress, cement houses, running water,
electricity, but what about development as equality for women? In our
family, in our municipality, if we do not have equality for women, then
there can be no development." During the rest of the workshop, the
diverse group of rural and urban women discussed issues of inequality,
domestic violence, sexual oppression, racial discrimination, the absence
of black women in public office, and other experiences affecting them as
Chocoan women.

On the following day, many of the women who attended the workshop assembled in Quibdó's central square to march in the International
Women's Day parade. While waiting for the band that was to lead the
parade to arrive, Torres led the group (including many women from
the cocina popular) in rally cries of "Mujeres organizadas jamas seran
explotadas (Organized women will never be exploited)" and "La mujer
que no protesta es una mujer sin vergüenza (The woman who does not
protest is a woman without pride)." Parra, one of the principal organizers of the parade, handed out coconut popsicles and led the gathered women and girls to celebrate black women with songs, poetry,
and jokes. When the band finally arrived, the women marched through
town amid much cheering, holding homemade banners. The messages
on the banners were written in colloquial terms and proclaimed the need
for attention to inequalities and abuses faced by women and celebrated
women's gender and ethnic roles.

Among the marchers were women from OBAPO. Their banners linked
women's concerns to Afro-Colombian struggles, listing the laws that
guaranteed the rights of women (Articles 40, 42, and 43) and of ethnic
groups (Article 7, AT 55, and Law 70) under the 1991 Constitution.

The PCN's position with respect to black women's struggles was similar to the one portrayed on OBAPO's banners in that it also saw gender
struggles as part of ethnic mobilizations. In an interview with Arturo

Escobar, PCN members expressed the need to engage gender issues—black women's roles and needs, issues of sexism and power inequalities, Afro-Colombian women's organizing—within spaces such as community councils (*consejos communitarios*) and the Black Women's Network (Organización de Comunidades Negras 1996: 256–60). They acknowledged that sexism and unequal relations between men and women were problems within black families and even in the black movements. But they believed that such issues should not be fodder for the "new gender politics," which they claimed would compromise gender and ethnic goals by encouraging women to split from ethnic movements and organize independently. In keeping with their broader ethno-cultural politics, the PCN was also critical of "institutional" approaches to gender, arguing that the state's development interventions exploit women's vulnerabilities and rely on the cultural principles of solidarity and support to insert black women into market cultures. These views were later echoed by Libia Grueso and Leyla Arroyo (2002). For the PCN, gender relations, being social and cultural relations, were part of the internal dynamics of the black movement:

> Many of us recognize that the relations between men and women are fundamentally cultural. This gender discourse as it has been unfolding is at the margins of social, political and cultural processes. The gender discourse as it has been raised by the majority of women's organizations is an "exogenous" one. It refers to the problems of black women but it does not take into account black cultural dynamics. . . .
>
> Still the social movement of the black communities must solve the gender problem. We can think of it as "a stone in the shoe" or as an opportunity, a possibility. This is an issue for the ethnic movement, not just a women's problem. (Organización de Comunidades Negras 1996: 259–260)

The PCN, especially its women leaders, perceived the Black Women's Network as an important vehicle through which to expand discussions of gender within the PCN and to gain more support for el proceso among black women (Grueso and Arroyo 2002). This was evident in remarks about the network made by its coordinating team (*equipo dinamizador*) at the January 1996 Workshop on Gender and Ethnicity (Balanta et al. 1997) and published in a special issue of the journal *Esteros*.[10] The equipo dinamizador, based in Buenaventura, consists of five members—Olivia Balanta, Betty Rodríguez, Sonia Sinisterra, Piedad Quiñonez, and Leyla Arroyo—all linked to the PCN. According to Balanta and colleagues (1997), it was the black mobilization process that played a pivotal role in strengthening the Black Women's Network and in focusing attention on gender and ethnic needs. The political and organizational principles of the network paralleled those of the PCN and emphasized autonomy from state interventions. According to their account, black women's efforts to organize in the region, including the dynamics in the cooperatives, were relegated to secondary status, after the ethnic struggles. Still, Betty Rodríguez (1997), one member of the equipo dinamizador, noted that the work of the network involved addressing a series of ambiguities and conflicts with and against the PCN, regional women's organizations, and Colombian feminists at large.

The dynamics and dilemmas of the shared ethnic and gender project in the Pacific were also the focus of my conversation with Fundemujer compañeras. Patricia Moreno noted,

> Historically speaking, the issue of gender has reached here recently. It arrived at the same time as the ethnic and regional issue. But the question is how to continue with the gender struggle without drowning or being subsumed by the proceso? How to help the ethnic struggle without giving up the gender struggle? Because of the *machista* culture here, women, too, recognize professionalism, but not womanhood. The women, too, forget their role.
>
> We all get lost in the hubbub of the Pacific (*generalidad del Pacífico*). One man in the PCN told me that they are afraid that if we talk about gender, we will forget the ethnic struggle.
>
> They are afraid that gender issues from the interior [the Andean center of the country] will dilute the ethnic struggle. They believe that we do not feel our own oppression. It is outside forces that "research" us—as if we did not feel our sexual, physical subordination.

But while the compañeras concurred with several elements of the PCN's perspective on gender, they had a different understanding of the network's origins and how to work toward their shared goals.

> *Myrna Rosa Rodríguez:* As black women, we also hold radical positions but differ from the "closed" radicalism of the PCN regarding who are our allies. We do not want to replicate mestiza women's mistakes. . . .
>
> *Dora Alonso:* I'm not with el proceso, but I believe in the struggle. I am involved in it. The PCN is very closed and jealous—jealous in the sense that if one is not physically "black," then one could not enter their struggle. I don't face the same problem within Fundemujer. I am part of the Red de Mujeres Negras del Pacífico. [The] PCN's radical position cannot even recognize or resolve their internal differences. I am not "black," but when

I support black issues, they look down on me. But my husband and my children are black. PCN folks are "sectists" [laughs as she tells me this is a pun on "sexist"].

Mercedes Segura: After much discussion, we called ourselves the Red de Mujeres Negras, but does that exclude mestiza women? No. Dora is a mestiza but also a "black," not because she has a black husband, but because she cares; she gets involved. One could be white like Gabriella Castellano [the director of the Center for the Study of Women, Gender and Society at UniValle] but also be "black" because she did not want to be superior to anyone.

Patricia Moreno: We want to learn from mestiza women's struggles and black struggles. We do not want to militate against anybody; nor do we want to be appropriated.

Like members of the network's equipo dinamizador, the Fundemujer compañeras saw the Black Women's Network as the place from which black women could collectively and autonomously organize to address their ethnic and gender concerns. But autonomy for Fundemujer meant making alliances with other entities, including the state, NGOs, and the PCN. It also meant—and here Fundemujer leaders were quite clear—retaining organizational independence from them all. Similar perceptions about black women's struggles prevailed farther south in Guapi, where members of the local network saw themselves as autonomous black women working to uphold Afro-Colombian traditions and to meet their "practical and material needs."

"Working from the Head Out": Black Women's Self-Revalidation and the Rescuing of Black Identity

In October 1995, I met Teófila Betancourt, an eloquent member of the Guapi Committee of the Black Women's Network. We had heard about each other and finally connected at the Fundación Habla/Scribe offices in Cali. We spoke for several hours about ethno-cultural movements and women's struggles. According to Betancourt:

Black women have helped Law 70 through their womanhood (*ser mujer*). However, they have little or no visible public or political role *as* black women. This is also true in black women's cooperatives. Black women are discriminated against triply—as poor people, as women, and as blacks. This leads them to undervalue themselves. Black women on the coast as well as in the cities are straightening, coloring, or perming their hair to "whiten" themselves and adopt customs of mainstream Colombian society in an attempt to gain social acceptance. This is a loss of value and beauty.

Like PCN activists, Betancourt considered the revalidation of their identity as black women to be an integral first task in the network's broader aims of helping Afro-Colombian women and their families, "so Network members organized hairdressing and hairstyling competitions (*encuentros y concursos de peinado*) among black women in Guapi. Now more than half the Guapireñas sport cornrows, braids, and other traditional Afro-Colombian hairdos. We need to work from the "head out" to regain our external identity as black women, to reflect on and affirm ourselves as *negras*, as we are with our own culture."[11] Betancourt said that as a CoopMujeres member, she both benefited from and helped with its work among Guapireñas. But having come from a small riverine *vereda* (district), she knew there was a need to address economic and social concerns among women living in remote rural areas of Cauca. Because these communities were beyond the reach of CoopMujeres, she began to organize rural black women living along the extensive network of Caucan rivers through the network's Guapi Committee. Betancourt invited me to visit Cauca and learn more about the network's work.

Four years later, in July 1999, I went to Guapi and spoke to Betancourt and her compañeras, among them Yolanda García, Luz Marina Cuero, and a young woman named Eden.[12] The compañeras also shared with me several published and unpublished manuscripts about and by the Guapi Committee, which had expanded to embrace seventy-four river-based and regional women's groups from Cauca. Later that evening, they showed me a videotape of the second meeting of women from Cauca and Nariño held in Timbiquí, Cauca, in May 1997.

The compañeras told me how a few self-help groups—Promoción de la Mujer from Río Saija, Grupo de Apoyo a la Mujer from Río Timbiquí, Fundación Chiyangua from Río Guapi—had emerged and organized around their chores: subsistence farming, collecting shellfish, making *guarapo* and *biche* (types of liquor) from sugarcane, taking goods for sale to the market. These groups organized to obtain better economic returns for their traditional activities and to share their experiences. Luisa, a woman from the Grupo María Auxiliadora del Río Saija, describes the work of black women in a copla (poem or popular song mostly composed and sung by women in the Chocó region) called *Trabajo de la Caña* (Sugarcane Work):

Son las cinco de la mañana	It is five A.M.
Me levanto a cocinar	I wake up to cook
Voy a filar mi machete	I will sharpen my machete
Para irme a trabajar	And go to work.
Coro	*Chorus*
Ay! Pobre mujer	Ay! Poor woman
Qué bonita estás (bis)	How lovely you are (repeat)
Llegando al cañaveral	On reaching the sugarcane field
Me encomienzo a rociar	I begin to weed
Y recojo mi cañita	And collect my cane
Para irla a cargar.	To take it with me.
Yo le digo a mi comadre	I ask my girl friend
Que me venga ayudar	To come work with me
Llamando nuestro hijos	Calling our children
Y mi espos dónde está?	But my husband, where is he?
Lo mando a cortar la leña	I send him to cut firewood
Pal' guarapo cocinar	To make guarapo
Ay! Marido, yo le digo	Ay! Husband, I tell him
Al tonel lo vamos a echar	We will put it in the barrel.
Ya el guarapo está fuerte	Now that the guarapo is strong
Lo vamos a destilar	We will distill it
Voy a arreglar mi cochito	I am going to fix my vessel
Y el viche sacando ya.	And we will brew viche.
A mi comadre una botella	A bottle for my girl friend
Que me vino ayudar	Who came to help me
Y el resto que nos queda	Whatever is left
Lo vamos a negociar	We will sell.
Este es nuesto proceso	This is our process
Para el vice sacar	To distill viche
Si ustedes lo quieren	If you want it
Pal' Saija a trabajar.	Come to Saija to work.

According to members of the Fundación Chiyangua:

> The organizing force of black women comes from work, from life itself. When women go in search of *chocolatillo* [a plant used in basket making] or to fish, they often leave for up to five days. Others stay in the house. For example, ones who are pregnant or unwell stay back with all the children. If I go, I leave the children with the neighbor, and she takes care of them. If I have an older daughter, she takes care of all the children, including the neighbors' children. That is the tradition, and it becomes a form of work. (El Hilero 1998: 16)

The groups from Río Guapi promoted the use of plants from their *azoteas* and *patios*, and developed menus of traditional dishes; the Río Timbiquí groups raised pigs and chickens for food and sale; the groups in Río Saija focused on extracting traditional products from local food crops (such as molasses from sugarcane). Others formulated projects to build houses, promote primary health care, find better transportation to and from regional markets, and obtain basic education for black women and their children. According to Teófila, each group engaged in activities that arose from its members' "needs, perceptions, and experiences as black women, and represented their collective strength."

To symbolize this strength, in 1996 the Guapi Committee of the Black Women's Network adopted an elaborate name: the Red de Organizaciones Femeninas del Pacífico Caucano Matamba y Guasá: Fuerza y Convocatoria de la Mujer del Pacífico Caucano (The Matamba y Guasá Network of Women's Organizations of the Cauca Coast: Strength and Convocation of the Woman of the Pacific Coast). The *matamba* is a very strong vine; the guasá, as noted previously, is a musical instrument traditionally played by women. The compañeras told me that Matamba y Guasá organized several successful events not long after its formation. Laughingly, they recalled the Timbiquí gathering in 1997:

> We had money from the Proyecto BioPacífico to organize a meeting for about sixty women. But 160 people turned up! Proyecto BioPacífico was worried about housing and feeding so many people. But we accommodated everyone in hotels, homes, [and] schools, and shared the adventure and chaos.
>
> It was great to see so many women together. Many had never left their veredas before, and now they were there to share their views and feelings, to speak. We discussed our rights, our identities; we cooked, we ate, we had a parade and danced—just us women!
>
> At one point when we all finished eating, we said that the washing up ought to be done by the men. The rural women were horrified at the idea of asking

the men to do the washing up. But we explained that *this* was all part of asking for women's rights—the rights of women in everyday life.

The Guapi Committee members noted that they neither think of themselves as subordinate nor consider their work antagonistic to that of their menfolk. Yet Eden said that their organizing "took a lot of work, especially for those who have men in their household." Of the seventy-four groups linked to the network, seventy were for "women only." In fact, men were welcome in these groups, noted the compañeras, as long as they "behaved themselves." While men were allowed to participate in group activities and meetings, only women made decisions.

Matamba y Guasá's efforts were aided through an Ecofondo–Proyecto BioPacífico grant for a project called Ríos Vivos (Living Rivers or Rivers Alive), which Fundación Habla/Scribe had helped facilitate. Fundación Habla/Scribe had long-term experience working in the region, including in remote areas, and believed that groups like Matamba y Guasá knew their realities best. In keeping with their overall philosophy, they supported local organizing and mobilizing in pragmatic and imaginative ways. Through the Ríos Vivos project, they helped Matamba y Guasá obtain funding for events such as the Timbiquí gathering and produce printed and audiovisual materials about their activities (Proyecto Ríos Vivos 2000).

Given the importance of orality and wordplay in Afro-Colombian culture, much organizing and mobilizing is done through sharing complaints, news, and experiences in the form of stories, songs, poetry, and coplas (such as the one by Luisa). Fundación Habla/Scribe helped collect and disseminate these and other statements and interviews about Matamba y Guasá.

As it grew, Matamba y Guasá drew increasing attention from state agencies and NGOs, who began supporting its efforts and calling on the organization to participate in conservation and development projects—and in the implementation of Law 70. Funding from the Proyecto BioPacífico, for example, supported "productive" projects, such as cultivation and recuperation of native food crops and medicinal plants, as well as "conservation" projects that focused on agricultural and silvicultural practices to conserve the environment and biodiversity (Asher 2004).[13] Network members also became active in community councils and worked with Ecofondo and the RSS on territorial zoning: demarcating collective property boundaries, drafting community-development plans, and outlining local efforts to preserve ethnic diversity.

I asked the compañeras what they thought about claims that, by engaging with state programs, their efforts would be co-opted or their ethnic struggles would be weakened. They insisted that "most of our ideas come from within the groups, not from the outside" and that only a few of the Matamba y Guasá efforts (such as its work with territorial zoning) depended on external funding. Members of the Fundación Chiyangua noted in an interview,

> This is another regional interest that unites those of us women who are working with it: the appropriation of territorial and ethnic rights as recognized in Law 70 of 1993. If anyone lives in constant relation with the land, it's women. The collective titling of land and socioeconomic development are two themes of much interest to grassroots organizations of the region, not just women, but also to men and young people, because we are defining what we want to do now to guarantee the future of the communities. (El Hilero 1998: 17)

They considered the network a place to bring together "women defending their ethnic and territorial rights, and working for the welfare of their families and their communities" (Red Matamba y Guasá 1997), and from which to establish alliances with broader cultural, political, and ecological processes in the region. Such alliances were not without tensions. As a member of the Fundación Chiyangua, Betancourt participated in state-sponsored "ethno-science" workshops. While she spoke positively of the information and knowledge exchanged during these workshops, she expressed skepticism about the utility of national "biodiversity databanks" for local communities: "We do not trust too many institutions and agencies. We speak with you because we know you especially through Fundación Habla/Scribe. But we prefer not to get involved in things we do not understand or with people and groups that we do not care about." With respect to their relations with the PCN, another Matamba y Guasá member noted:

> We work or help the process from our ser mujer. We have our own methodology, our own way of doing things. We do not help them formally in the [PCN], but there are some things that we do in parallel, and at other times we are involved in different issues with different methods and different objectives.
>
> We meet them in public spaces, but we maintain our characteristics. We interact and reach a consensus, but we do not want to get involved in clientelist networks. Rather than obtaining representation and power, we look for spaces of participation for black women.

Addressing my question about how they dealt with the differences among them, the compañeras said that they were aware of the differences but that they had chosen to identify and ally around commonalities and to reject the notion that black women are primarily "victims" who need aid from outside. These issues are underscored in the vision and political perspective of the Black Women's Network:

> It is important to clarify that the meeting spaces [of the network] are generated and constructed by us, with our own initiatives. We have been struggling for recognition of women in our region and to overcome [the obstacles to recognition]. Activities such as ours imply sacrifices, imply surrender to make our dreams come true and to achieve the proposed objectives. Beginning from these principles today we are ready to identify ourselves as women and come together as a gender, to recognize our similarities and differences.
>
> We do not want to be represented by anyone. We want to be considered protagonists of our lives and of our world. (Red Matamba y Guasá 1997–1998: 15–16)

Though not quite as visible as Matamba y Guasá, several more locally based black women's networks in the Pacific were also actively organizing around ethnic identity and gender concerns in the post-Law 70 period. Such activism involved Afro-Colombian women asserting the primacy of ethnicity to demand that the state address their concerns as women and as blacks. By invoking their experiences and perceptions as black *women*, they simultaneously engaged and went beyond struggles to obtain black ethnic and territorial rights.

Reflecting on the energy and determination of black women's organizing, it is easy to romanticize their gender and ethnic subjectivity and agency. But I want to caution as much against such romantic assertions of Afro-Colombian women's organizing and resistance as against devaluations that proclaim it co-opted by the state and market forces.

Reading Afro-Colombian Activism: Differences, Discourses, and Dialogues

Issues of women's needs and gender identity are central to the discourses of the Black Women's Network of the Pacific. The network's claims about identity, autonomy, self-representation, relations to land, and politics emerging from everyday experience closely parallel the ethno-

cultural discourses of the wider black movement. Indeed, while not naming groups such as Matamba y Guasá specifically, Grueso and Arroyo
(2002) celebrate black women's activism as clear examples of anti-
globalization, where women contest outside impositions from *"lo propio
(their own perspective)."* They write, "Place-based women's struggles in
the Pacific confront the market-led model of neoliberal globalization by
opposing the accompanying cultures of consumption, food dependence,
and erosion of autonomy in other basic elements of well-being" (Grueso
and Arroyo 2002: 70). At the same time, the language of development
and environment that frames the new political-economic changes in
the region also constitutes "black women" as a development category
(Álvarez 2000). Such language is actively appropriated by the women
themselves. A copla presented during the second working meeting of
Matamba and Guasá in Timbiquí in 1997 is a case in point:

BIODIVERSIDAD

La Biodiversidad es parte
De nuestra preocupación
Y por eso en el evento
Fue un tema de atención

Y en los tiempos de ahora
Nadie se puede quejar
Pues organizaditas
Mucho podemos lograr

Capacitación queremos
En técnica agropecuaria
Formación social
Y lideres comunitarios

Fue un tema de atención
Para invitarlos a todos
A seguir con los cultivos
En los tiempos de ahora

Las mujeres de estas tierras
Nos estamos preocupando
Por el progreso, el cultivo
Y por seguir organizando

BIODIVERSITY

Biodiversity is part
Of our concern
That is why it was
A theme of our meeting

In these times
We cannot complain
Because organized
We can gain much

We want to train ourselves
In agricultural technologies
Social formation
And community leadership

It was an important theme
To invite all
To continue with our plantings
In these times

We women of these lands
Are concerned
About progress, about culture
And about continuing to organize

Es un trabajo muy duro	It's very hard work
No hay quien diga no es verdad	Nobody can deny it
Con sacrificios y esfuerzos	With sacrifices and strength
Se llega a un buen final	We can reach a good end

Written (or, mostly likely, sung first) in traditional form, the copla engages recently arrived concepts such as "biodiversity," "training," "agricultural technologies," and "progress" that form the basis of mainstream economic development. The juxtaposition of conservation, development, and women's sacrifices alert us to the limitations of considering afrocolombianas' organizing merely as resistance or empowerment. Indeed, women are becoming participants and decision makers in economic and cultural processes that affect them and their communities. But through their use of and engagement with the language and institutions of development interventions, they help produce such discourses. The organizational structures in women's networks are also increasingly bureaucratic, with secretaries, treasurers, regional coordinators, and other "officers" coordinating the groups' efforts.

One could see such engagement with development and gender interventions as "an appropriation of the material and cultural resources of local struggles," as PCN activists feared. But celebrations of black women's place-based activism by the PCN through formulations such as Grueso and Arroyo (2002) could just as well be seen as appropriation of black women's gender struggles by ethnic movements.

Conceptualizing black women's activism (or, indeed, ethno-cultural struggles) in binary terms—autonomous assertions of alternatives to globalization on the one hand and subject constitution by state and development discourses on the other—is not helpful, precisely because it fails to capture the complexities of how local movements become vehicles for interventions such as those of the market and the state. Discourses of gender, women's rights, and revalidation of black ethnic and cultural identity are used in the Pacific to contest the hegemonic notions of state-sponsored development. At the same time, black women's subjectivities (as with Afro-Colombian agency in general) are shaped through and against political-economic interventions. Ethno-cultural politics complicates but does not negate this relationship.

This does not mean relegating the network's engagement with "institutional" interventions and outside entities to the category of "cooptation." Such an explanation underestimates the importance of work involved in building networks across both similarities and differences—

work that black women regularly undertake. It also risks viewing these alliances as a natural affinity rather than a politics of solidarity. Since the late 1990s, the political and economic realities of the Pacific lowlands (as of other parts of Colombia) have taken a turn for the worse. There is an increasing presence of armed forces (guerrilla, paramilitary, and military) and drug dealers in the region. As I discuss in the next chapter, local communities get caught in the crossfire of the escalating violence; their lands become battlefields without warning or recourse; and they face massive and repeated involuntary displacement from their homes. Given the inability of the Colombian state to ensure the basic human rights of its citizens in the context of such complex political realities, viewing black women's activism as either autonomous organizing or co-optation by the state is too simplistic.

Going beyond the limits of mainstream liberal (struggle for equality) and marxist (struggle against class-based exploitation) approaches, feminist scholars explore how Latin American women's political participation is shaped by a variety of factors including capitalism, colonialism, nationalism, racism, and sexism (Alvarez 1990; Jaquette 1994; Jelin 1990; Safa 1990; Schild 1998; Westwood and Radcliffe 1993). The experience of Afro-Latino women (see "Mujeres negras, Latinoamerica" 1995), including that of Afro-Colombian women (see Camacho 2004; "Mujeres Negras e Indigenas Definiendo Futuro" 1997), is key to understanding the dynamics of both women's activism and broader cultural and political movements.

Local movements engage development interventions in paradoxical ways. Based on her work on women's movements in Ecuador, Amy Lind (2005:19–20) emphasizes "the political and economic *paradoxes* of women's struggles for survival, as a way to illustrate how women and men, in Ecuador and throughout Latin America, interpret and negotiate ideas about development, state modernization, globalization, and modernity in complex ways." While being constituted as a "development category," black women simultaneously channeled state attention and resources to their identity and needs as *black* women (Álvarez 2000).

Pondering a development policy for and by women in the Colombian Pacific, Maria del Rosario Mina (1995) writes, "We have a basic interest: to work together among ourselves and with others to propose a politics of development for Afro-Pacific women that corresponds with our basic existence (*ser*), from which it is possible to identify the material and social conditions and needs (*tener*), that guarantee that existence." For Mina, a focus on material needs is not at odds with recovery of, and pride in, identity. In her formulation, ser (existing) and tener (having that which guarantees existence) are linked. For black women to participate in state projects is not just to forsake their identity and traditions to obtain the things that development promises. On the contrary, according to Mina, to engage with development is to engage in the struggle for power to define what is understood by "needs" and to devise means to meet them. This requires women to engage strategically with development interventions as well as with ethnic struggles. Even as they are constituted as development's subjects in the political and economic agenda of the state, afrocolombianas and the "new gender politics," are disrupting cultural relations as they have been constructed. To cite a single example, the PCN now explicitly and increasingly engages women and gender concerns (Flórez-Flórez 2004).

Black women's concerns and the responses of afrocolombianas' networks emerged within the context of multifaceted, intertwined, and mutually constitutive relations of power—of gender (as women), of race or culture (as blacks), of class (as poor people), and of location (as rural, Pacific residents) at a time when blacks were granted special rights and black social movements were in the process of "ethnicization" to translate new laws into concrete results. Predictably, each struggle affected the others: meeting basic needs remained a central concern of black women's cooperatives because prevalent economic models destroyed or failed to provide adequate livelihoods. Nor had the broader black struggles yet provided concrete economic alternatives; productive activities therefore remained central to women's concerns. At the same time, as local leaders of the network such as Sylveria Rodríguez and Teófila Betancourt noted, these political-economic concerns were linked to the marginalization and exploitation of Afro-Colombians. That afrocolombianas were aware of the multiplicity and complexities of these realities and were negotiating them skillfully is evident from the astute remarks of network members. Stressing that the primary aim of the Black Women's Network was "revindication of ethnicity, gender and appropriation of territory," black women's groups attempted to strengthen the organizational links between and among black communities and establish political alliances beyond the region (El Hilero 1998; Red Matamba y Guasá 1997, 1997–1998). Afrocolombianas were also keenly aware of their audiences, knowing well what could and could not be articulated within the fractured political economy and cultural politics of the Pacific.

The example of black women's cooperatives and the Black Women's Network highlights how afrocolombianas' subjectivities and organizations are shaped by the language and practices of development institutions and their gender politics, as well as by broader black cultural and political movements. In turn, black women drew on the prevalent terms and resources of development initiatives and regional cultural discourses to address their needs and reshape the agendas of both the state and ethnic struggles. Through such mutual but uneven constitution, they helped to institutionalize state interventions but also channeled state resources and discourses to address their ethnic and gender needs. Black women in the Pacific region linked their ser and tener both rhetorically and practically, to speak about their experiences, analyze them, and propose ways to overcome their marginalization. In the post-Law 70 period, Afro-Colombian women's strategies were shaped through their active engagement with and against the development practices of the Colombian state and black ethnic movements.

5

DISPLACEMENT, DEVELOPMENT,
AND AFRO-COLOMBIAN MOVEMENTS

This book ends the way it begins: by drawing attention to how conjunctures, contradictions, and contingencies accompany the simultaneous unfolding of development and black social movements in the Pacific lowlands of Colombia. But if the adoption of the new constitution in 1991 and of Law 70 in 1993 struck some as the dawn of a more peaceful era for the country and of a new hope for Afro-Colombians, the end of 1995 seemed to mark the twilight of those hopes.

As my year of fieldwork was drawing to a close in December 1995, the government was once more facing charges of corruption. Headlines screamed of the capture of the Arjuela brothers, who headed the Cali drug cartel, and the scandal over the Samper administration's links with the cartel. Corruption was one of many elements that plunged the country into yet another political and economic crisis.[1] Another was the rapid spread of guerrilla and paramilitary forces, including in the Pacific. This was a new factor in the region, and one that proved definitive in years to come. Since 1993, my many trips between Cali and Buenaventura, and to the Anchicayá River and areas south, had been without incident. After October 1995, it became common to be stopped by uniformed men for "security checks." Travel along the Cali–Buenaventura road was considered dangerous even in daylight hours, especially along the Dagua track, which was a known guerrilla stronghold.

Less than a decade after adopting a new constitution, Colombia faced another impasse. After a brief and abortive attempt at peace talk with insurgents, the official response became one of stepping up the military offensive (against guerrillas and drug traffickers) and economic apertura, a policy maintained to this day. Both efforts have the financial and political backing of the United States. Social, environmental, and cultural concerns have remained on the development agenda but are subsidiary to, or subsets of, military confrontation and economic globalization.

Since the late 1990s, local communities in the Pacific increasingly have been caught in the crossfire or have become specific targets of violence, death, and displacement. As black communities literally and figuratively "lost ground" in the Pacific, Afro-Colombian movements faced new splits and splinters. On the one hand, the mid-1990s saw a proliferation of local community councils (*consejos comunitarios*) formulating proposals for collective land titles. These local-level councils often had no connections to regional black organizations such as the PCN or ACIA, or they kept their distance from them. However, with the acceleration of armed violence and the targeted killing of leaders and activists, local councils could not exert control over their lands even if they had managed to obtain collective titles.

On the other hand, the number of black communities displaced by violence began to take on alarming proportions. This led the three black sectors described in the first part of this book (the PCN, Chocoan groups, and black politicians) to form coalitions across their differences and with other social causes in the nation to address the problems of displaced Afro-Colombians. Increasingly, these sectors have also rallied international support, connecting with human-rights entities, churches, solidarity groups, and African American politicians. As Afro-Colombian movements expanded to connect with black groups beyond Colombia, the terms "Afro-Latino" and "Afro-descendant" gained wider currency. The call for recognition of black identity and territorial rights also began to be reformulated to stress the right of black communities to return to their homes and collective lands to live lives of dignity and peace. As displaced Afro-Colombians face new forms of discrimination and invisibility, there is a resurgence in activism against racial discrimination, as well as for socioeconomic and political equality and reparations for indignities suffered in the distant past and in more recent times.

A comprehensive exploration of all of these changes would be the start of another book. My purpose in this final chapter is to discuss the changes in development and black movements in relation to the projects, people, and places that are the focus of previous chapters.

The twilight of Afro-Colombian hopes, like their dawn, is tinged with various hues. The new strategies and configurations of black social

movements are no less heterogeneous, contradictory, or intertwined
with state-led changes than those that emerged at the beginning of the
Law 70 process. Given these new conjunctures and contingencies, the
issue of establishing critical solidarity with black struggles without representing Afro-Colombians solely as victims or, alternatively, romanticizing resistance remains a crucial one.

Revisiting Development, the Environment, and Ethnic Rights in the Pacific Lowlands

After 1995, the latent national crisis and low-intensity conflict in Colombia emerged fully into the foreground and accelerated. The collapse
of the Cali drug cartel led to the disintegration of many sectors of the
regional economy. In the summer of 1999, I saw stark evidence of economic decline in Cali: abandoned construction projects, closed shops,
unmaintained streets and public spaces, and a dramatic increase in the
number of migrants and refugees. There was also a palpable presence
of the armed forces, with military helicopters hovering over the Cauca
Valley and along the Buenaventura road. This was a direct response to
the increased presence and activity of insurgents. The hijacking of an
Avianca plane and the taking of forty-one hostages (including U.S. na-
tionals) and the kidnapping of 160 people at a Cali church in May 1999
were but two of the audacious guerrilla actions that received international attention.

President Andrés Pastrana responded to the crisis with yet another
set of peace talks with the largest guerrilla groups, FARC and the ELN.
These talks were opposed by large landowners, agribusiness interests,
and certain elements of the middle class. As they turned to drug trafficking, extortion, and kidnapping to fund their operations, guerrilla forces
were also losing legitimacy among other Colombian sectors. Consequently, the progress of the talks was uneven, with military and paramilitary forces often attacking guerrillas in the declared demilitarized
zone in the south. Not surprisingly, perhaps, this set of negotiations was
short-lived.

Its replacement was the U.S.-funded Plan Colombia, which started
as an anti-drug campaign but soon expanded into an armed offensive
against guerrillas. In 2000, the Clinton administration approved an initial $1.3 billion aid package for Colombia.[2] This aid mainly took two
forms: the first included herbicides, helicopters, and spraying equipment

to aerially fumigate and destroy coca crops; and the second included weapons, military equipment, intelligence, personnel, and training for the Colombian army and police.[3] A small component, less than a fourth of the total amount, was earmarked for "alternative development projects" to promote the cultivation and marketing of non-narcotic crops. Most of the fumigation and drug-clearance operations were concentrated in the Putumayo and Caqueta regions of southern Colombia, but the ripple effect was felt in the Pacific states, especially Nariño.

As a drug-control and eradication strategy, mechanized aerial spraying has proved ineffective. The first few years of the campaign saw an actual increase in the amount of land under coca cultivation; subsequent decreases have been insignificant. Military operations have been equally ineffective in wiping out guerrilla insurgents. Rather, these anti-narcotic and anti-insurgency tactics fueled conflicts between drug traffickers, guerrillas, paramilitary forces, and the Colombian army to gain military and economic control over the region.

When he assumed office in 2002, President Álvaro Uribe stepped up military operations sharply, aiming to recover guerrilla-controlled lands.[4] As one U.S. post-Cold War crusade (the "war on drugs,") morphs into another (the "war on terror"), Uribe's tactics have enjoyed strong political and financial backing from President George W. Bush. The putative rise of the left in Latin America adds Cold War dimensions to these "narco-guerrilla" wars, with Hugo Chávez's popularity in next-door Venezuela particularly raising U.S. fears of revolutionary movements spreading in the Andes. According to a brief by the Council on Hemispheric Affairs (COHA) in 2006, Colombia had received $4.5 billion in foreign aid since 1997 and become the third largest recipient of U.S. foreign aid.[5] This aid has done little to achieve the goals of "peace and security" (or what President Uribe calls "*seguridad democrática*," or "democratic security"), which remain ever illusive for the Colombian masses. The Uribe government, like its predecessors, has been charged with corruption and links to paramilitary forces. The government and the army stand accused of disdaining democratic processes, abusing human rights, and generating another crisis of justice in the country.

The military and economic strategies that prevail in Colombia are also steadily eroding the environmental and ethnic gains of the 1990s in the Pacific. Forests and natural ecosystems, as well as cultivated food crops, have become the "collateral damage" of the non-selective chemical herbicides used to eradicate drugs.[6] Areas cleared of natural vegetation and coca are quickly replaced with more coca or taken over for agro-industrial

gh-profit, export-oriented enterprises dominate the agro-
ojects, among them plantations of African oil palms, which
ystems and subsistence crops. Black and indigenous com-
at the center of this vortex, facing death and forced dis-
t goes almost without saying that their ability to enforce
ethnic and territorial rights is another casualty.[7]

Figures on the total numbers of internally displaced persons differ,
with the Colombian government's figures lower than those of major
human-rights groups. The Consultancy on Human Rights and Displacement (Consultoría para los Derechos Humanos y Desplazamiento;
CODHES), one of the most respected and authoritative nongovernmental sources on internally displaced persons in Colombia, estimates that
between 1995 and 2005, 3 million people were forced to flee their homes
because of violence related to armed struggles or disputes over territory
and resources (Consultoría para los Derechos Humanos y el Desplazamiento and Pastoral Social 2006: 1). According to the International
Committee of the Red Cross, despite the supposed demobilization of
paramilitary groups in 2005, internal-displacement figures remain on the
rise: from 45,000 in 2005 to 67,000 in 2006 and an estimated 72,000 in
2007 (Comité Internacional de la Cruz Roja 2007). Internally displaced
persons tend to belong disproportionately to minority groups, with Afro-
Colombians accounting for around 33 percent of the 412,500 persons
displaced during 2002 (Human Rights Watch 2005). During the first
quarter of 2006, 12 percent of the total number of those displaced were
from minority groups (Consultoría para los Derechos Humanos y el
Desplazamiento 2006: 3).

That the Pacific region could become a theater of conflict and conten-
tion over land and resources was foreseen by Afro-Colombian activists
and the state. Since the start of the constitutional-reform process, these
entities have engaged in legal and political battles to determine the future shape of the region and its natural-resource wealth. The struggle
to pass and implement Law 70 was part of Afro-Colombians' attempts
to insure their future by obtaining territorial rights and consolidating
their political and organizational ability to manage and administer their
lands autonomously. For its part, the Colombian state focused on re-
organizing the regional economic-environmental and ethno-territorial
bases through instruments such as *ordenamiento territorial* (territorial
zoning). But neither development projects nor Afro-Colombian movements had consolidated their strength when the latent conflict in the
country erupted in full force across the Pacific region.

By the turn of the twenty-first century, then, "violence" and "displacement" had entered the vocabulary of the Pacific alongside such now familiar terms as "biodiversity," "natural-resource wealth," "poverty," "economic isolation," and "blackness." In the second half of the 1990s, however, Plan Pacífico and related projects such as Proyecto BioPacífico remained the official development guides. These plans continued to emphasize sustainable natural-resource-based development, integrating the regional economy into national and international markets while paying attention to the social, cultural, and ethnic rights. When Proyecto BioPacífico ended in 1998, its mandates were passed on to the Quibdó-based John von Neumann Institute for Pacific Environmental Research (IIAP).

After three years of discussions to define its statutes, the IIAP began operations in 1997. The IIAP's principal aim is to coordinate scientific research and development in the Pacific and ensure that they benefit local communities. This aim is governed by the politics of territorial zoning and many of the ethno-territorial concepts outlined by indigenous and black communities at the meeting in Perico Negro in 1995 (see chapter 2). As a first step, the Ministry of the Environment charged the IIAP with outlining a politics of conservation and development for the region for the twenty-first century. From 1998 to 2000, the IIAP held discussions and meetings with diverse sectors involved in the Pacific, including government agencies such as the ministries of the Environment, Health, and Interior, the Land and Agrarian Reform Institute (INCORA), and the Social Solidarity Network (RSS); indigenous and black groups; regional development corporations; university programs; local chambers of commerce; and other business sectors. The outcome of these meetings was the document "Agenda Pacífico XXI: Proposal for Regional Action for the Diverse Pacific for the New Millennium."

I was unable to locate a full text of this document. However, in referring to "Agenda Pacífico XXI," Sander Carpay (2004) and Carlos Agudelo (2005) frequently cite terms such as "sustainable development," "alternatives to capitalist economic growth," "community-based resource management," "participative democratization," "autonomy," "decentralization," "empowerment," "green markets," "food security," and "ethnic-territories." While Agudelo is more critical than Carpay in his evaluation of "Agenda Pacífico XXI," both suggest that it is an excellent blueprint for a locally planned, community-managed, participatory, culturally sensitive, ecologically sustainable model for development and environmental conservation.

In practice, the IIAP has been unable to live up to the promises of "Agenda Pacífico XXI." Aside from questions of the compatibility of its multiple aims and the difficulties of executing such a broad spectrum of tasks, there are many reasons the IIAP failed to implement the agenda. The institute was a new institution whose objectives were contentious from the beginning (as was evident at the meeting to discuss the statutes described in chapter 2). It had little political or financial support from the national or regional governments. Neither did local government bureaucrats, community members, or NGOs in the Pacific consider it an important, or often even a legitimate, interlocutor (Agudelo 2005).

One reason for this disregard was the inexperience and lack of qualifications of the IIAP's personnel. Another was that many of the key positions in the IIAP were occupied by members of the Chocoan political elite and their cronies. In separate conversations in July 1999, Mauricio Pardo of ICANH and Alberto Gaona of Fundación Habla/Scribe told me that discussions around the election of the first director of the IIAP (Rudecindo Castro) were fraught with conflict. Gaona spoke of indigenous delegates walking out of a meeting in Bahia Solano because of the *politiquería* in the IIAP. These veteran activists from the PCN and women's cooperatives told me that many valuable opportunities to debate and decide the kinds of research needed in the Pacific were lost as the IIAP became another hotbed of Chocoan politicians.

This seemed to bear out in the case of Bismarck Chaverra (the Chocoan lawyer mentioned in the first chapter), who became the IIAP's director in 2003. Chaverra, whose family had political connections in Chocó, had been a vocal presence at the Ecopetrol audencia and the Buenos Aires meeting to discuss the Development Plan for Black Communities (PDCN). However, since those meetings there had been a visible distance between Chaverra and community activists in all fora I attended in 1995. Possible explanations include internal divisions and the differences between the political strategies of Chocoan organizations and the PCN. Whatever the cause, the rift meant that the IIAP did not have the weight of the black movements' leadership behind it in implementing "Agenda Pacífico XXI."

Many indigenous communities in the Pacific also distanced themselves from the IIAP. In March 2003, five indigenous organizations wrote a strongly worded letter that they circulated widely via email. Denouncing the IIAP for being overtaken by traditional politicians, the letter is critical of Chaverra's candidacy and the process by which he was elected director. Lamenting the breakdown of genuine interethnic dialogue and accusing the IIAP and the Ministry of the Environment for not respecting indigenous perspectives, the email from the leaders announced that

> the indigenous people of the Pacific and their organizations no longer consider themselves members of the IIAP. We leave its destiny in the hands of Afro-Colombian leaders and opt for the autonomy of our people, as we have always done regarding the management and care of our territory and the knowledge that emanates from them.
>
> This implies that the IIAP is prohibited from conducting research in our territory and is not considered a valid interlocutor between our communities and Afro-Colombian communities. (Carta Pueblos Indígenas Pacifico)

In his evaluation of "Agenda Pacífico XXI," Diego Cháves Martínez (2005: 90) concludes that seven years after its conception, the significant time and effort expended on it had had little impact. Most regrettable, in his opinion, was the institute's loss of credibility and the disappointment among local communities.

The accelerating violence, the subsequent displacement of peoples, and the reorientation of political energy have also contributed to the uneven implementation of the agenda (Agudelo 2005). While the IIAP's ability to carry through with its ambitious aims remains doubtful, it serves as a conduit for conservation and development projects from several regional offices in the Pacific. Various versions of "Agenda Pacífico XXI" also continue to circulate and exert their influence in maintaining the interlinked discourses of environmental conservation and sustainable development in the Pacific.

Meanwhile, the National Planning Department's policies for the Pacific region echo Plan Pacífico, stressing economic growth, infrastructure and institutional development, and meeting basic human needs. A draft of the 1998–2002 National Development Plan entitled "Changes for Peace" refers to the mandates of Law 70 and focuses on six areas to promote development in Afro-Colombian communities: land titling and acquisition, basic sanitation, health, education, productive activities, and institutional development. The 2002–2006 and 2006–2010 versions, "Hacia un estado comunitario (Toward a Communitarian State)" and "Estado Comunitario: Desarrollo para todos, su aplicación en el Pacífico (A Communitarian State: Development for All, Its Application in the Pacific)," respectively, also mention these issues and focus particularly on human rights and gender.[8]

On paper, long-term plans for "black/Afro-Colombian, *Palenquera* and *Raizal*" (from palenques such as San Basilio and the Caribbean islands of San Andrés and Providencia) populations are a visible part of government development policies. Environmental concerns, couched in terms stressing their economic and market value, are also part of the region's development agenda. Reference to "Agenda Pacífico XXI" also appears in the 2002–2006 draft in connection with ecological and territorial zoning. Indeed, environment, development, and territorial issues have become more and more intertwined, not just in the Pacific, but in the nation as a whole. In 2003, the Ministry of the Environment was restructured to become the Ministry of Environment, Housing, and Territorial Development. Unfortunately, this intertwining has not meant greater attention to social or ecological principles. On the contrary, as Cháves Martínez (2005) writes, past and present governments have given little actual support to projects with broad, long-term aims such as "Agenda Pacífico XXI." Development plans instead stress the need for official investment in high-profit, export-oriented commercial sectors such as oil and natural gas.

Ambitions to improve the well-being of the population through economic growth notwithstanding, development statistics for the Pacific remain grim in the short term. In a 2003 report assessing Colombia's performance in achieving the Millennium Development Goals, the Pacific and Afro-Colombian communities continue to appear at the bottom of the list of human-development indices, with 85 percent of their basic needs unsatisfied (Sarmiento Gómez et al. 2003). The report notes that "social expenditure per capita" in the Pacific departments (except in Valle del Cauca) is lower than in other parts of the country. According to data from the Observatory of Racial Discrimination at the University of the Andes, 76 percent of Afro-descendant people live in extreme poverty, 42 percent are unemployed, and only 2 percent reach university. Infant mortality among Afro-descendants is three times higher than among other children in the country: the average life span of a black child is three years (Alto Comisionado de las Naciones Unidas para los Refugiados 2007). These figures parallel those of the National Planning Department.

The most tangible development success in the Pacific appears in statistics for collective land titling. According to Enrique Sánchez and Roque Roldán (2002), under the Natural Resource Management Program funded by the World Bank (the PMRN project discussed in Chapter Two), 113,954 people (22,132 families organized in 58 community councils) were granted collective titles to more than 2.36 million hectares, over 20 percent of the Pacific region's 10 million hectares. Based on data from the Colombian Institute for Rural Development

(INCODER), which replaced the Colombian Institute for Agrarian Reform (INCORA), Agudelo (2005) estimates that by 2003, almost 5 million hectares of land had been processed, with 127 collective titles granted to 148 community councils. World Bank reports laud the Natural Resource Management Program's land-titling component for taking a participatory approach and claim that it has strengthened the ties between communities and state institutions, helping the former to face down the violence of paramilitary groups and guerrillas. However, such positive assessments of community–state linkages are neither universally shared nor borne out consistently by events on the ground.

Concomitant with the granting of the collective titles were threats, death, and displacement among black communities. The first lands to be collectively titled in 1997 were in the lower Atrato River. However, community members and leaders were not present to receive titles to more than 70,000 hectares because the violent events in the river basin in 1996 and 1997 had forced them to flee (Wouters 2001). This was a consequence of the spread to the Atrato region of paramilitary forces, including the United Self-Defense Groups of Colombia (Autodefensas Unidas de Colombia; AUC) from the Urabá and Sucio river regions in Antioquia and northern Chocó.⁹ Over the next five years, as guerrillas followed the paramilitaries into the region and the two groups battled for control, the communities of Vigía del Fuerte and Bellavista were caught in the crossfire. Each side accused local residents, especially community leaders, of sympathizing with the other (Agudelo 2005).

At the end of April 2002, the two armed factions were locked in serious battle, which ended in a tragic event that drew international attention. On May 2, the FARC set off a cylinder bomb inside a church in Bojayá, Chocó, sheltering three hundred people. The resulting explosion, which killed 119 people and injured 98, was among the most serious in the country and was globally condemned. Reports from various human-rights organizations, the Diocese of Quibdó, and the United Nations denounced not only guerrillas and paramilitary groups but also the Colombian government for not responding to warnings about this offensive that they had received days earlier. Continued threats and fears of such violent offensives led to an exodus of refugees from the region. Drawing from reports by the Life, Peace, and Justice Commission of the Diocese of Quibdó, Mieke Wouters (2001) notes that between 1996 and 1999, 417 assassinations were registered for the middle Atrato region; and that in the second half of 1999, eight thousand displaced people arrived in Quibdó.

Sometimes the violence was linked directly to issues of land titling. In the Cacarica River basin in the lower Atrato region, members of the community councils were subjected to threats, harassment, and displacement before they finished their formal applications for collective titles. In 1998, 120 community councils linked to ACIA in the middle Atrato region received titles to approximately 700,000 hectares. Soon after, many community members, especially *communitarios* and other leaders, faced threats, kidnappings, harassment, and murder (Wouters 2001). Besides the loss of life and livelihoods, such violence and forced displacement meant major setbacks for the communities' ability to come up with the required management plans for their lands.

The situation farther south was comparable. Since 1999, guerrilla and AUC forces had been a palpable presence in coastal Valle del Cauca, with the latter claiming to have arrived to "clean" the region of the former. In 2000, massacres in Sabaletas and other rural communities in the Buenaventura municipality led to a massive displacement of people to Buenaventura. In December 2001, Naka Mandinga, by then a leader of the community council of Río Yurumanguí, circulated an email message to solidarity groups typical of the communiques arriving from black and indigenous organizations. Mandinga wrote that collective titles covering 54,000 hectares along the Yurumanguí River had been granted on May 23, 2000. Less than a year later, in April 2001, paramilitary forces in the neighboring Río Naya accused one hundred fifty people (blacks, indigenous people, and mestizos) of being guerrilla collaborators and killed them. Armed men then entered the vereda of El Firme in Río Yurumanguí and killed seven members of the community council who were out fishing. More than one thousand people were displaced — some four hundred fifty to Buenaventura, and six hundred more along various parts of the river.

Subsequently, the remaining Yurumanguí council members regrouped in Buenaventura and organized the return of vereda residents. But paramilitary forces returned in December, threatening to disrupt Christmas festivities unless the council members left the river. They also promised to kill the council members' kin in Buenaventura unless they were obeyed. Similar threats of massacres and murders were made in Río Naya, Río Raposo, and Río Cajambre.

In June 2000, CODHES declared a temporary alert in thirty-seven veredas of the five principal rivers (the Anchicayá, Naya, Raposo, Yurumanguí, and Cajambre) in the municipality of Buenaventura. According to CODHES and other human-rights sources, about 60 percent (approximating 20,000 people) of the rural population around Buenaventura was at high risk of being displaced.

Agudelo's (2005) overview of the situation in coastal Nariño presents an equally grave picture. Paramilitary groups in Tumaco started issuing "black lists" of guerrilla sympathizers in 1999 and "cleaning up" those deemed undesirable. According to official records, paramilitaries assassinated seventy-six individuals between September 2000 and May 2001. Guerrilla forces soon followed, taking control of various municipalities on the Nariño coast.

Nor were drug traffickers to be outdone. In 1995, there was only a smattering of coca fields in coastal Nariño. Over the next ten years, this increased to 20,000 hectares. The Diocese of Tumaco, a longtime supporter of black and popular organizations, was a vocal critic of the human-rights violations committed against local communities by all three extralegal forces. On September 19, 2001, in Tumaco, a pair of gunmen linked to the paramilitaries killed Yolanda Cerón, a nun and director of the Catholic church organization Pastoral Social in Tumaco. Church and secular human-rights agencies such as Amnesty International were among those that led the protest against Cerón's assassination.

Paramilitary and guerrilla actions did not lead to as many mass displacements in coastal Cauca as in the other parts of the region.[10] It was for this reason that, in 1999, I was more easily able to visit Guapi and follow up on the work of black women's groups than in other parts of the coast. Yet the new military posts in Guapi increased incidents of illegal resource extraction and violence in the southern Pacific were beginning the spiral of death and displacement, a process that only intensified under President Uribe's Plan Patriota.

Death and displacement are not the only kinds of violence suffered in the region. Communities that return to their veredas and rivers find their livelihoods obliterated or severely threatened because of the direct and indirect destruction of forests, natural ecosystems, and cultivated crops. At the end of the 1990s, the logging firm Maderas del Darién reappeared in the Cacarica River basin in the Chocó and renewed its operations in lands titled to local communities and in surrounding areas. The firm's old permits, which had been revoked when ACIA began its struggles for territorial rights, had apparently been reissued by CODECHOCO.

Making a dubious situation even more so was the fact that the renewal occurred while the Cacarica communities were displaced from their lands after paramilitary offensives in 2001. In response, the Cacarica Communities for Self-Determination, Life, and Dignity (CAVIDA) filed suit against CODECHOCO, the Ministry of the Environment, and Maderas del Darién for violating Law 70 and various environmental accords.[11] Even while the case was waiting to be heard, the firm continued its operations. According to messages circulated by black activists, company personnel gave chainsaws and mechanical towing equipment to local community members, urging them to "log in the name of development."

In the southern Pacific, aerial fumigation rather than illegal logging functions in conjunction with paramilitary pressure to clear land of vegetation and people. Roundup, the chemical herbicide made by Monsanto, is the most commonly used under Plan Colombia to destroy illicit crops, and it is non-selective. Glyphosate, the active ingredient in Roundup, kills all vegetation—food crops, herbs, trees, and shrubs—with which it comes in contact. Nor is aerial fumigation as target-specific as officially maintained. Studies by various nongovernmental entities show that it causes deforestation, destroys subsistence food plots, contaminates soil and water, kills aquatic life, and makes humans and animals ill (Peterson 2002). Local testimony collected by solidarity groups, journalists, and NGOs also document the negative effects of chemical fumigation on ecosystems and human health. Typical of such testimony is a message describing a fumigation event written by the PCN and the community councils of Río Timbiquí, Cauca, and circulated electronically by Arturo Escobar:

> On May 20, 2006 at 3 p.m., four small fumigation planes and five helicopters flew over the middle part of the Timbiquí River between the communities of El Charco, Mataco, and San Miguel del Río. The fumigation lasted for approximately an hour and a half, but it was sufficient to devastate important subsistence crops such as *papachina, chivo, banano, yuca*, and *plátano*. That the fumigation in San Miguel had grave consequences is evidenced by illnesses and blindness suffered by those who were sprayed in the face by the liquid (chemical). The football field was completely burnt. Children, pregnant women, old people and the whole community were bathed in a strange liquid. The only thing that is known about this liquid is that it leaves black men and women without food to feed their children.

Officials deny such reports or dismiss them as support for the insurgency. In contrast, independent assessments of aerial eradication strategies show that chemical spraying of coca plots displaces drug cultivation from one area to another, often pushing it deeper into forests, including national parks and nature reserves. These areas were initially off-limits to fumigation. However, despite much controversy and public outcry, after 2004 the Uribe government authorized fumigation in protected areas.

Another threat to the environment, and to ethnic groups in the region, comes from land speculators and investors turning to "frontier" areas. Large tracts of land are being bought in the southern Pacific by private companies and agribusiness elites. Lands cleared of vegetation and people by illegal resource extraction or anti-narcotic measures, or through the territorial dispossession of ethnic groups by private speculators, are replanted with coca or used for agro-industrial schemes.

Most egregious of these are the plantations of African oil palm mentioned earlier. Oil extracted from these palms is used as a commercial solvent in industry and as a bio-fuel. African oil palms were introduced in the Pacific through development schemes in the 1960s, but plantations covering thousands of hectares are a recent phenomenon. Environmental studies confirm that such monocultures lower the water table, leaching nutrients from the already nutrient-poor tropical soils and contaminating them.

Links between paramilitary forces and agribusiness interests have been confirmed by various sources, including the United Nations High Commissioner for Refugees (UNHCR), the Colombian Attorney-General's Office (Procuradoría General de la Nación), the Human Rights Protection Office (Defensoría del Pueblo), Catholic church organizations, journalists, and ethnic and environmental groups (Richani 2002; Salinas 2005; Censat Agua Viva and Proceso de Comunidades Negras 2007; Mingorance et al. 2004; Roa Avendaño 2007). Citing data from IN-CODER, Salinas (2005) notes that 10,000 hectares of oil palms are located in the collective lands in the Jiguamiandó and Curvaradó basins of the Chocó and extend to the Cacarica basin. Nevertheless, these activities continue, often with the blessing of regional development corporations such as Corpourabá and CODECHOCO.

The increasing global interest in bio-fuels as an alternative to crude oil and petroleum-based energy sources makes African oil palms a potentially lucrative investment. Private speculators with links to the military and foreign capital (especially oil and energy companies) depend on paramilitary forces to protect their economic interests (Richani 2002). Such agro-industrial development was anticipated in Plan Pacífico, and stimulating bio-diesel production is a key part of President Uribe's economic strategy for the region. Since 2003, the cultivation of African oil palms has received increasing state encouragement and funding in the form of credits and loans. Giving demobilized paramilitary groups land and encouraging them (and black communities) to engage in agro-industrial development is also related to Uribe's "peace and resettlement" plans. These plans have support and funding from the U.S. Agency for International Development (USAID), as African oil palms are considered a commercially viable alternative to coca, notwithstanding their problematic social and environmental effects. Fedepalma, the association of owners of African oil palm plantations, aims to expand production to 1 million hectares in Nariño, and the government has proposed that by 2020, 7 million hectares in the department will be used for oil palms and other export crops (Bacon 2007). Bacon and Roa Avendaño (2007) cite President Uribe's address at a Fedepalma congress in June 2006, where he noted:

> I strongly request [the Secretary of Agriculture] to lock up the business community of Tumaco together with our Afro-descendant compatriots and not let them leave the room, keep them there until they come to an agreement. There is no other choice. . . . Lock them up and propose that they . . . that the state . . . come to an agreement about the use of land, and the government will supply venture capital. And give them a deadline and tell them: Gentleman, we are in session, and we will not leave here until we have an agreement . . . because we must recognize the good and the bad. Here in Meta and in Casanare and with what's beginning in Guaviare, we've had powerful palm tree growth, but none in Tumaco. And Tumaco, with its highway, and little to the north, the Guapi area, El Charco has excellent conditions but not one palm tree, just coca, which we need to eradicate.

To expand oil palm plantations, President Uribe created incentives (such as tax breaks on bio-fuel production) and imposed higher taxes on regular fuel. Despite opposition from a broad sector of Colombian society and state entities, including the attorney-general and a few parliamentarians, he also reformed agrarian policies and natural-resource-management laws to stimulate agro-industrial development. These same Colombian sectors, including environmental, indigenous, Afro-Colombian, and peasant organizations, strongly criticize the new forestry, water, and mining laws (adopted between 2001 and 2006) for giving short shrift to environmental conservation and social goals (Arias 2006; Colmenares n.d.: Guhl Nannetti 2006; Vélez 2005). With respect to the new natural-resource-management legislation, former Minister of the Environment Ernesto Guhl Nannetti (2006) wrote:

They are applying an extractivist, antiquated and destructive model
prevalent from the 19th to the middle of the 20th century rather
of sustainability and social inclusion as is appropriate for the 21st century.
Furthermore, the old models served more as means of territorial appropriation through violence and stimulation of poverty and marginality rather than
providing access to forest resources.

Adopted in the name of "sustainable development," but underscoring
the spirit of neoliberal economic growth, these laws are seen as laying
the groundwork for a U.S.–Colombia free-trade agreement. In an interview with an independent news source, Carlos Rosero (2007) noted:

> Territory as a life chance and political chance is something that is under im-
> mense threat not just because of the war, or the mega [development] projects,
> but also because of the legislative measures of the Colombian government,
> such as the Forestry Law, the Statute regarding Rural Development, the min-
> ing code reform that is about to be adopted of the mining code that is about
> to happen, and the Water law. To us, the blacks, indigenous and peasants,
> these [new legislative measures] are a sign and a call that say "Hey! Careful!
> The wolf is coming; the wolf is here." The wolf is the state, those associated
> with the state are the wolves—the firms, the multinationals that want to live
> off of what is not theirs.

Indeed, the restructuring of the Ministry of the Environment and IN-
CORA was part of the process to formulate and implement these new laws
in accordance with Uribe's favored principles of economic growth.

In September 2005, I spoke to Claudia Mejía and Camila Moreno in
Bogotá about their perceptions of the changes in the Pacific. Both had
worked for several years with different state entities, including the Human Rights Protection Office (Defensoría del Pueblo), the organization
responsible for promoting and protecting human rights in Colombia.
Moreno had gone on to work with a Swedish NGO, while Mejía worked
with Social Action (Acción Social), the federal welfare program that
replaced the Social Solidarity Network. Reflecting on the situation in
the Pacific, both said that economic strategies represented by the radical expansion in the cultivation of African oil palms signaled a return to
the "extractivist" approaches of the pre-1991 Constitution period. They
felt that the 1991 Constitution had signaled a moment of real hope and
change for the country. Ordinary Colombians and ethnic groups had begun to have a stronger sense of their citizenship, and their rights and new
environmental and sociocultural policies were influencing and giving

a sustainable shape to development. They believed that these foci, especially as they applied to the Pacific region, were real and not merely rhetorical. As such, both were disappointed that more than a decade later there had been no significant change in the political and economic situation. Moreno noted that the influx of illegal activities and armed forces in the region were less a cause than a consequence of a weak state.

In contrast to Moreno, Mejía was cautiously optimistic about the future of the region, noting that state programs and the land-titling process funded by the World Bank could help protect territorial rights and provide services for displaced populations until the current crisis subsided. But Mejía agreed with Moreno that local communities had little real control over their physical and political spaces.

I met Moreno and Mejía again in August 2007, separately this time. Moreno was working with the Swedish Embassy in its program on human rights and humanitarian assistance. While reiterating that the situation in the Pacific and for displaced Afro-Colombians remained grim, Moreno was inclined to be positive about the work of Colombia's Constitutional Court. Moreno told me that the court had been taking various steps to ensure that the state actually provided services for displaced communities and implemented the laws designed to ensure their basic rights.

Mejía was still at Social Action, working on the Proyecto Protección de Territorios y Patrimonio de la Población Desplazada, a project to protect the territory and property of displaced populations. Mejía located this special territory-protection project within the context of Law 387 of 1997, which defines internally displaced people and guarantees them services and protection.[12] As the state welfare agency, Social Action is charged with implementing Law 387 and overseeing service provision for internally displaced people. Law 387 had just had its tenth anniversary, and although non-white ethnic groups disproportionately figured among those displaced, the law did not (and as of this writing, does not) have specific components to address their concerns. Neither does it have mechanisms to protect the territories and property of those who are forced to leave their homes.

The land-protection project was initiated in 2002 to address the latter issue and in response to a legal decree in 2001 governing the property rights of displaced populations. The project's mandate was to ensure that people forced to flee their lands did not lose them permanently. Taking the "right to property" as a fundamental human right, the Social Action project team had begun by assessing the problem, developing methods to define "territorial entities" that needed protection, and devising strategies to prevent the abandoned lands from being usurped by third parties.[13] However, the mechanisms, or "networks of protections," conceptualized under this project were for individuals and individual property rights and did not take into account the territorial rights of ethnic groups and the issue of collectively held land. The land protection project team had begun grappling with the issue of ethnic lands only in 2005, and thoughts on adjusting existing mechanisms to the circumstances of ethnic territories remained preliminary.

Perhaps in response to my impressed but somewhat incredulous expression, Mejía observed dryly that the project arose not from state benevolence but from the fact that, as a member of an international community and a signatory of international human-rights accords, and as a recipient of international aid, Colombia was required to comply with these accords, implement its own laws, and ensure the basic human rights of its citizens. Initial funds for the project came from sources such as the Post-Conflict Program of the World Bank, the UNHCR, and the International Organization for Migration. In its later phase, the project received funding from Canada and the European Union. The Colombian state contributed few financial resources in the early phases but eventually is expected to generate the funds to implement and sustain the project without international support.

As Mejía described it, the project sounded admirable but largely composed of bureaucratic and institutional instruments of land protection. I asked Mejía whether it worked, or was likely to work, "on the ground." Mejía grimaced as she said that, as with many other Colombian mechanisms, the laws and policies for internally displaced people were excellent. However, she admitted that neither the institutional structures nor sufficient political will existed to guarantee rights. She was especially pessimistic about the latest rural-development law, Law 1152 of 2007, and its effects on the territorial rights of black communities and environmental protection in the Pacific. Law 1152 had been passed on July 25, 2007, just a few weeks before our conversation. Mejía claimed that all territorial-protection mechanisms conceptualized prior to the rural development law, including under Law 70 of 1993, had been subsumed under the new law. Prior to the passing of Law 1152, the territorial rights of ethnic groups had been governed by the Office of Ethnic Affairs in the Ministry of the Interior (formerly the Division for Black Community Affairs) and INCODER. But in 2008, control over these collective lands and over all territorial units in the Pacific passed on to a new entity called the

"Land Unit (Unidad de Tierras)," which has not yet been established. This law, like many recent agrarian, natural-resource, and environmental laws, is a controversial one because it is seen as paving the legal path for the aforementioned U.S.–Colombia free-trade agreement. According to Mejía and many critics, ethnic and environmental issues are a small—almost an insignificant—part of these new accords. With so much of the Pacific area under the law's jurisdiction, Mejía felt she had cause for concern as "over 1 million hectares [in the Pacific] are in the process and under the land protection project. But many, many communities still do not have collective titles, and others have not even managed to put in proposals to get them. But this controversial law and this new institution are going to take over land titling." Shaking her head ruefully, Mejía reiterated her doubts about the possibility of adequately protecting either ethnic territories or the Colombian environment under the measures taken by the Uribe regime.

Activists and professionals with both long-term and recent links to the Pacific echo Moreno's and Mejía's sentiments. Since the late 1990s, many of the old environmental and developmental NGOs—Fundación Habla/Scribe, Fundación Herencia Verde, FES—have disbanded or have been severely restructured because of decreases in social funding. In September 2005, Alberto Gaona and Alberto Valdés, formerly with Fundación Habla/Scribe, told me that while some of the old activists and professionals went on to work with other organizations, there was little possibility or space for them to work toward broad social change within the current administration's push for agro-industrial development led by the private sector. The two men remarked bitterly that the new militarized and neoliberal pathways to "security" and "democracy" (referring to President Uribe's policy of "democratic security") are dismantling progressive policies and laws of the 1990s. Within this new conjuncture, social activists and professionals within NGOs and state organizations are forced into a rearguard action, mitigating the rollback of gains made in past struggles. Gaona spoke from firsthand experience about state organizations, as he had worked as a consultant for the Human Rights Protection Office with Mejía and Moreno.

Felipe García of the Equilibrium Foundation, an environmental NGO, told me in August 2007, "It is only possible to work on concrete projects and things (*cosas y proyectos puntuales*) in the Pacific." The general consensus in 2007 is that radical or alternative possibilities for ethnic rights and social movements in the Pacific region are increasingly circumscribed by the violent tendencies of neoliberal economic globalization and the armed conflict. Political and cultural rights are increasingly couched in terms of "human rights" rather than demands for socioeconomic equality, recognition of difference, or justice for citizens within a democratic nation-state. This change in the discourses of social and ethno-cultural struggle occurs in conjunction with a proliferation of international agencies, especially those providing humanitarian assistance and relief services, recording human-rights violations, and accompanying communities to prevent displacement or enable their return.[14]

Shifts in Post-Law 70 Black Movements: From Afro-Colombians to Afro-Descendants

As stated elsewhere in this book, Afro-Colombians long foresaw the probable arrival of violent conflict in the Pacific region. Their struggle for ethnic and territorial rights was in part pre-emptive: they sought to be in the strongest position possible when and if such conflict erupted. Black activists, especially in the PCN, attempted to rally different Afro-Colombian groups and organize a social movement around the principles of "identity, autonomy and territory." As PCN spokespeople noted, a strong black movement was necessary for many reasons, including the need to prevent the displacement and disenfranchisement that might lie ahead (and did). But as discussed in chapter 3, building a unified black movement across the heterogeneity of Afro-Colombian identities and interests posed challenges that were difficult to overcome, especially in the short term.

The challenges, contradictions, and differences that Afro-Colombian movements faced took on new forms in the late 1990s, as waves of violence and forced migration from the Pacific generated new forms of invisibility and marginalization among black communities. One atrocious problem is that Afro-Colombians displaced as a result of coca eradication are not considered internally displaced persons under Law 387 of 1997 (the law defining the government's obligations to internally displaced persons) and therefore are not eligible for protection.[15] Whatever their official displacement status, large numbers of Afro-Colombians arrive in Andean centers, often after suffering multiple displacements (from rural communities to various regional centers and from there to major cities). The state's laws regarding displaced people, in many respects among the world's most advanced, fail by dint of a seeming technicality to reach some of the most tragically displaced people in Colombia.

As their circumstances change, so do the forms of racism Afro-Colombians must face. The anthropologist Marta Pabón began working with the Afro-Colombian Subdivision of the Office of Ethnic Affairs in early 2007. I met her on August 14, 2007, and asked for her assessment of the Afro-Colombian situation and what was being done to address it through her office. Among other things, Pabón told me about her first evaluation visits to the Pacific region in May 2007 and what she described as discrimination against displaced Afro-descendant people in Tumaco, Nariño.

On her arrival, Pabón was told of a displacement event that had caused seven thousand people to flee from nearby El Charco. Two officials from the regional branch of Social Action took her to see thirty displaced families who were living in a church in Tumaco. Conditions in the church were crowded and unpleasant. One family was huddled in a corner, with a child sleeping on the floor, even though there was a pile of mattresses close by. The Social Action official pointed to the scene and said that this was a classic case of black ignorance and neglect. The official said that it was typical of a black mother to allow her child to sleep on the floor, even though mattresses were available. At this, Pabón went to the woman and asked why the child was sleeping on the bare floor rather than on the mattresses. The woman, who was indeed the child's mother, said that during the night the family had been fiercely bitten by the bedbugs and fleas that infested the filthy mattresses. After that the family had decided to sleep on the floor or on mats rather than receive bites that might infect.

Later that day, when Pabón recounted the incident to another Social Action employee, he said that he and his office were overstretched and did not have enough resources to attend to the number of people who needed attention. According to Pabón, when this official had been approached earlier by one of the displaced families with an application to register them officially, he responded, "You Afros have big families and networks of relations. You don't need our aid."

Such stories abound. Black activists, journalists, and human-rights observers note that displaced Afro-Colombians are regularly denied services, despite the substantial institutions created expressly to guarantee the rights of displaced citizens.

Having escaped one set of violent forces in rural areas, Afro-Colombians arriving in the crowded slums of Andean cities face others: urban gangs, militias composed of demobilized guerrilla or paramilitary groups, drug pushers, prostitution rings. Luz Marina Becerra of the Association of Displaced Afro-Colombians (Asociación de Afrocolombianos Desplazados; AFRODES) recounted a story she heard from a young black woman who had recently arrived in Bogotá. The young woman told Becerra that being displaced felt like running from one set of dangers into another. She had been displaced from a riverine community in the Chocó to a regional center. The arrival of paramilitary groups made her flee again, to Quibdó, and from there to Bogotá. Far from feeling safe in Bogotá, she felt like she had to run from danger again, and the dangers in Bogotá felt worse because they were multiple and unknown. Added to the material dangers were those that came from feeling alienated from her cultural and physical environments.

Black sectors first organized in the early 1990s began to reorient and change their strategies toward the end of the decade. Leyla Andrea Arroyo and Víctor Guevara of the PCN, for example, became affiliated with an evangelical church and worked with the displaced population in Buenaventura through projects funded by the church and various state entities, such as the Human Rights Protection Office and the Social Solidarity Network. Yellen Aguilar continued to focus on ethnic and environmental issues through work with the World Wildlife Fund and other environmental NGOs. In addition to organizing black communities, PCN and Chocoan activists began to form or renew alliances with other political sectors (including traditional political parties) and social groups in Colombia and abroad. They also began to rally international support for their ethnic and territorial rights, drawing attention to the new forms of invisibility, marginalization, and racism faced by Afro-descendants. PCN activists were at the forefront of these efforts.

In the late 1990s, many of the PCN's key activists started networking outside the Pacific region and the country. Such outreach had always existed to a certain extent, as when Arroyo traveled across Europe in November 1995 to obtain financial and political support for black struggles. Now, with the acceleration of the Colombian conflict, ethnic, economic, and environmental issues in the Pacific began to gain even more international attention.

Concerns for the cultural rights and traditional knowledge of ethnic minorities had also been a focus of attention within development policy and activist circles for the past decade. With Law 70 considered one of the strongest pieces of legislation for Afro-descendant groups in Latin America, Afro-Colombians were invited abroad to share their experiences. Speakers such as Libia Grueso, an active spokeswoman for Afro-Colombian concerns, attended various anti-globalization protests in Latin America, Indians, and Europe.

These travels provided PCN activists opportunities to obtain support for Afro-Colombian struggles and to make connections with other movements, transnational groups, and policy advocates. In addition to connecting with NGOs and social activists, PCN intellectuals and professionals strengthened their affinities and connections with academics in the United States and Europe. In November 1998, Rosero and Grueso gave several talks at Bates College (where I taught at the time) and spoke at a Latin American studies conference at the University of Massachusetts, Amherst, before visiting several other universities in the United States. At Bates and at the University of Massachusetts, Rosero and Grueso spoke about "identity, territory and autonomy" for Afro-Colombians and other aspects of the PCN's positions, as outlined in chapter 3. Locating black struggles within the political-economic and geopolitical context of the Pacific and of Colombia, they expressed fears that cultural and environmental concerns in the Pacific would be subsumed by the violence of macroeconomic development and armed conflict.

I heard Grueso give a similar presentation in Washington, D.C., at the 2001 meeting of the Latin American Studies Association, at which she became a regular participant. In 2002, Arturo Escobar (with support from many others, including myself) nominated the PCN for the Goldman Environmental Prize for Latin America. The organization did not receive the award in 2002, but in 2004 the award was given to Libia Grueso. The nomination had been for Grueso, Rosero, and Cortés, but the Goldman Prize Committee preferred to recognize a movement through an individual and chose Grueso. This prestigious award increased the visibility of the PCN, of Grueso, and of their vision of Afro-Colombian struggles, especially on the international stage. In addition to this transnational activism, Grueso continued to work with the various state entities—the Division of National Parks, the Ministry of Culture, and, in 2006, the National Planning Department as a consultant on development plans for black communities in the Pacific.

In March 2004, I visited Colombia after a hiatus of five years and met up with Carlos Rosero, who had returned from a year of exile in Ecuador, to discuss the situation of black struggles and what had been going on in the Pacific in the intervening years. The meeting, in which other members of the PCN (Mario Angulo, Rolando Caicedo, Alfonso Cassiani) participated, occurred at Cassiani's house in Cali. I had last met Cassiani and his girlfriend, Julia Congolly, in Cartagena in 1995. Cassiani and Congolly, now married, had moved to Cali to work with black struggles from there. Congolly was on her way out of the house to work with Grueso on a report on a PCN project. She told me a little about the project, which focused on fostering productive activities among women of the Mayorquin and Yurumanguí rivers in Valle del Cauca.

I remarked to Rosero that Congolly's project seemed a departure from the kind of activities the PCN had pursued in the past. Rosero agreed and noted that in the first ten years of its existence, the PCN had made many advances and learned much from its mistakes. Both he and Cassiani agreed that black movements, including the PCN, needed to take stock of their past work as they looked ahead and strategized for the next phase of the struggle. They spoke of their desire to do this through a meeting of PCN members and the supporters and critics who had researched them or worked with them in some way. Rosero noted that among the lessons learned were the importance of addressing the immediate needs of poor black communities in rural areas and to organize urban black communities in places such as Buenaventura and Cali.

The black movement was in a different place than it had been in the early 1990s, and the PCN needed a new vision. While calls for "identity, territory, and autonomy" and collective land rights in the Pacific remained necessary, they were not sufficient focal points for organizing black communities. With growing numbers of Afro-Colombians displaced and forced to migrate to regional centers or Andean cities, their plight had also acquired an urgency that black movements needed to engage.

Both Rosero and Cassiani believed that working across black differences and with other sectors of society was more important than ever. In an essay published in 2002, Rosero offered similar reflections, noting that Afro-descendants

> should join forces to overcome the profound political and organizational dispersion, the absence of proposals and common action which defines the current situation of their organizations. Although strength must be found through hard battles even against Afro-descendants themselves, it must also be realized that in the context of the critical reality of the country . . . the future depends on the capacity of Afro-descendants to connect with the struggles

and aspirations of other excluded and subordinated social sectors and create strength by linking together weaknesses. (Rosero 2002: 558)

That the PCN was trying to work with other political and social sectors was evident later in the day when I accompanied Rosero and Rolando Caicedo, a young activist and a councilman (*consejal*) from Buenaventura, to visit Angelino Garzón, then the governor of Valle del Cauca. Although the national vote in 2002 had gone to the right-wing Uribe, many regional and municipal posts all over the country had been won by independent or left-leaning candidates. Garzón was affiliated with the Polo Democrático, a new, independent left-wing political party. When we finally met with Garzón, Rosero explained that he wanted to discuss both broad and specific concerns with him, referring to the issue of black territorial rights and how they were to be addressed in local, municipal, and regional development plans. Garzón responded positively to the importance of meeting the medical and other needs of black communities in Buenaventura, especially the people displaced there from rural areas.

As I discussed in the section on territorial zoning at the end of chapter 2, many entities, including multiple state agencies, are involved in implementing territorial rights and working with black communities to draft natural-resource-management plans for their collective lands. However, the mandates and responsibilities of these many parties are frequently unclear or conflicting, as is the case with jurisdiction over resource management and economic development on black lands. Garzón agreed that the political and economic crisis in the Pacific had complicated this already unclear state of affairs. Garzón invited black communities and their organizations to bring up the issue of black territorial rights and economic development on their lands at a public meeting (*audencia publica*) to discuss the plan for the Department of Valle del Cauca. He said that, in "the spirit of democratic participation," it was absolutely necessary for him to take into account black concerns, as it was for black communities to participate actively and positively in public discussions of economic development in the state.

With respect to the more concrete concerns, Garzón advised Caicedo to connect with the nurses' and doctors' unions, the mayor's office, politicians, and other organized groups that were grappling with the social, economic, and political crises in Buenaventura. Picking up the thread, Caicedo mentioned a forthcoming meeting of black Colombian mayors to be held in Buenaventura in May and expressed his hope that Gárzon would offer the mayors his support. Rosero and Caicedo sought support from Garzón on another specific issue: ensuring that there was an accurate count of the black population in the 2005 census. According to Rosero, the PCN and other black organizations had been working on a mass-media and education campaign to address the issue of the low levels of black "auto-identification." Garzón answered that there were laws that required local media to help facilitate citizen participation for the public good. He said that they all needed to mobilize to ensure that the 2005 census that included an accurate count of Afro-descendants was a success. The meeting ended on a cordial note.

Rosero's remarks and the meeting with Garzón suggest that the PCN's position with respect to formal or institutional politics had begun to change. Broader Afro-Colombian involvement in traditional politics had also been changing. In 1996, Afro-Colombians lost their special seats in the Chamber of Representatives on another technicality. Despite this setback, both Agustín Valencia and Zulia Mena stood for elections again in 1998 but lost. The former returned to his political base of the Conservative Party in Valle del Cauca, and the latter returned to regional organizing in the Chocó.

Other black candidates fared no better in the 1998 national elections. The exception was Piedad Córdoba, who was reelected to the Senate. Córdoba had been involved in the process to pass Law 70 and obtained a good percentage of the black vote. However, Agudelo (2005) notes that, although she mentioned black ethnic issues, her 1998 campaign strategy was firmly based on her traditional Liberal Party platform.

By the 2002 elections, the black quota of two seats in the Chamber of Representatives had been reinstated. This time, Carlos Rosero, Zulia Mena, and Piedad Córdoba joined ranks at the eleventh hour and campaigned together. Rosero, the PCN, and Mena, through her organization Black Peoples of Colombia (Pueblo Negro de Colombia), lauded Córdoba for her contributions to advancing black struggles and supported her candidacy to the Senate. Mena and Gabino Hernández (a lawyer and PCN member from Cartagena) also stood as candidates for the Chamber of Representatives under the special circumscription for blacks. Córdoba was successful in her bid for reelection to the Senate, thanks to her Liberal Party base, but Mena and Hernández lost the elections to two Afro-Colombian sports stars: Willington Ortiz, a football (soccer) player from Nariño, and María Isabel Urrutia Ocoro, a weight lifter and Olympic gold medalist from Cali. Neither Ortiz nor Urrutia had any prior links to black struggles, and their campaigns did not emphasize black identity or rights. During their terms in office, they did little to address the socioeconomic inequality of the vast majority of blacks in the country, and especially in the Pacific region.[16]

In 2006, Rosero made an unsuccessful bid for the Chamber of Representatives. The seats were won by Urrutia and Antonio José Caicedo Abadía, who was affiliated with a little-known organization called Raíces Negras. In her 2006 reelection campaign, Córdoba reiterated her commitment to "social multiculturalism" and made a stronger appeal to work for the betterment of black communities. Córdoba was reelected senator for the 2006–2010 term and is a strong critic of the Uribe government, militarization of the country, rampant violation of human rights, and unregulated free trade. Córdoba also articulates a strong antiracist and anti-segregationist stand—but again, from within the boundaries of affiliation with the Liberal Party.

Since the 1998 elections, a proliferation of black candidates, including from among black grassroots organizations, have stood for various offices at the departmental (assemblies, governors, chambers) and municipal (mayors and councilmen) levels. But at these tiers, as well, the black electorate's turnout has been low. Agudelo (2005) asserts that the situation reflects a lack of interest in electoral politics among black communities. He notes what black grassroots activists have observed: that traditional politics is still viewed with suspicion among the majority of black Colombians. This is partly explained by the (historical and actual) marginalization of blacks in political office, as well as by the pervasive corruption in Colombian party politics. Black voters, especially in rural areas, also have little expectation that their concerns will be understood and addressed through the channels of party politics or formal political institutions in far-off capital cities.

Peter Wade (1993a) attributes the low interest in electoral politics among blacks to the notion of the "ethic of equality." With specific reference to politics in the Chocó, Wade notes that given black history of discrimination and marginalization, Afro-Colombians tend to disavow any hierarchy among blacks. Further and more systematic research on the Afro-Colombian electorate and black participation in traditional politics is needed to follow up on these insights. However, they do bear out in anecdotes. In August 2004, following up on the activities of the women of Matamba y Guasá (see chapter 4), Alberto Gaona told me that Teófila Betancourt had stood unsuccessfully for the post of councilwoman in Guapi in 2002. According to Gaona, her political aspirations led many women from rural, riverine areas to distance themselves from her.

Among black social movements, ambiguity regarding traditional politics intensified after the election of Uribe in 2002 and the entrenchment of his right-wing policies with his reelection in 2006. On the one hand, black activists were more skeptical and pessimistic than ever about the possibility of bringing about progressive social change through mainstream political channels. On the other, they felt it was more imperative than ever to draw political attention to, and keep it on, the struggles, rights, and needs of their peoples.

When I met with Rosero again in August 2007, he said that Afro-descendant groups were pursuing these goals through alliances with other activists. Rosero noted that many members of the PCN and other black movements were now associated with the Polo Democrático, which had put forth a presidential candidate in the 2006 election. He said that there were many friends and allies of the black movements in the Polo and that he engaged with them regularly. He went on to tell me that, at their next meeting, he intended to propose that they organize a forum on the problems of the black population as a societal problem in Colombia rather than the concern of but one sector of the national populace. It was time, he claimed, to link up with non-black allies to address the issues facing marginalized communities across the country.

With increasing U.S. involvement and funding for the problematic Plan Colombia, black movements were also making connections with solidarity groups, Washington think tanks, and development NGOs. Also proliferating were links with U.S. politicians, especially the Congressional Black Caucus. Representatives of these entities visit Colombia regularly and work in various ways to keep Afro-Colombian issues on the U.S. policy agenda. Instrumental in fostering these links were activists who regularly traveled abroad, such as Mena and Rosero; politicians such as Piedad Córdoba; and displaced Afro-Colombians living in exile in the United States. Among these exiles are Luis Gilberto Murillo (former governor of the Department of Chocó) and Marino Córdoba (an activist from Chocó). Like many activists, Córdoba left the Pacific in 1996 after receiving death threats. In 1999, with PCN members and others, he established AFRODES. In 2002, continuing threats forced Córdoba to leave Colombia, but he continued AFRODES activities from the United States.

On August 14, 2007, in Bogotá, I spoke to two key AFRODES activists, Jattan Mazzot Ilele and Geilerre Romana. Mazzot began by giving an overview of AFRODES's aims and activities. He said that AFRODES tries to bring together Law 70 and Law 387 (the latter governing the rights of displaced populations in Colombia). AFRODES is meant to serve as an interlocutor between the state, the human-rights establishment, and black communities. Mazzot noted that members wanted to emphasize that "we are black groups, not just displaced peoples. We conducted a socio-demographic study to analyze the causes of displacement, to show that it is a systematic process that affects black communities. It is not a random occurrence."

Black cultural identity, socioeconomic development, organizational strengthening, and equity and participation—AFRODES's four thematic axes—and its slogan ("Territory, Culture, Autonomy, and Life") parallel the PCN's early organizing. The association works in conjunction with international human-rights and policy groups, such as the Washington Office on Latin America (WOLA), and national NGOs such as CODHES to facilitate the provision of education, health, housing and other services for displaced blacks. Through its alliances with the Congressional Black Caucus, AFRODES has also received USAID funds to promote alternative development projects in the Pacific region. In addition to these initiatives, Mazzot noted that, as a "recognized part of Colombian civil society," AFRODES participated in the United Nations World Conference against Racism in Durban, South Africa, in 2001.

The "right to return" is a central principle of AFRODES's work. Mazzot observed that when displaced individuals arrive in cities, they lose touch with their cultural identity and connections. The process affects young people with particular intensity, he added. Part of AFRODES's mandate is to ensure that Afro-Colombians retain their sense of black identity, their *ser*. "We work to celebrate and show who we are, through games, dance, music and theatre," said Mazzot. AFRODES has main offices in Cartagena, Quibdó, and Buenaventura and smaller ones in Cali, Tumaco, and Barranquilla. Through third-party alliances, their work also reaches communities in Neiva and Medellín.

Asked to comment on the National Planning Department's draft socioeconomic-development plan for black communities, Mazzot said, "It's a palliative and not generated in a participatory way." Geilerre Romana elaborated:

> The theme of forced internal displacement is a complex one, but we are more than [internally displaced persons]. We have legal collective titles over 5 million acres in the Pacific. There's more land to be titled. Although we have legal titles to our lands, in effect they are being appropriated by violent paralegal and extralegal forces. We need guarantees over our territory so we can re-

turn and live in peace and enjoy appropriate development. At the same time, we have to be realistic and develop a strategic, political vision. Those displaced from Bojaya [Chocó] also tried to return but without any guarantee that they can re-create their conditions of life. The issue of return is complex. But how do we live as an ethnic group without a territory? We fear we will disappear into society. Therefore, we need public policies that will recognize our difference and pressure the state to consider our need for appropriate development.

Romana also addressed the irony of food insecurity among black communities living in a resource-rich region. He lamented that inadequate social spending (13 percent of the national budget) resulted in the Colombian government's being unable to meet the "minimal guarantee of the state—food security." Romana reiterated that, despite numerous laws, constitutional guarantees (including a recent Constitutional Court ruling asserting that the government had to fulfill its obligations to meet the needs of Afro-Colombian internally displaced people), and international pressures, the Colombian state was not taking real action either to meet the needs of displaced Afro-descendants or to guarantee their ethnic and territorial rights. Indeed, the president denied the very existence of racism in Colombia. Here Romana was referring to President Uribe's response to critiques of his minimal attention to Afro-Colombians and how they were being disproportionately affected by violence. But then Romana continued,

> It is difficult to address the issue of displacement because of lack of opportunity, without a politics of peace, without a politics of the state, without social investment. . . . Still, since 1993 there have been regular national conferences on Afro-Colombians. They provide a space for the heterogeneous peoples who are Afro-Colombian and the plurality of movements to interact. In 2002, there were six hundred organizations that participated. We came up with a navigation chart to help us address the many elements of an Afro-Colombian project. AFRODES is giving visibility to these important tasks and achieving solidarity among black groups nationally and abroad.

Romana reiterated Mazzot's earlier remarks: "Law 70 is too rural in focus, and that does not help us address the issues facing the growing Afro-urban population." But he acknowledged that, while the situation was grave, it would be graver without the law.

Indeed, Law 70 has led to many gains for black groups. In addition to collective land titles and reserved seats for black delegates in the Chamber of Representatives, special entities such as the Afro-Colombian Subdivision in the Office of Ethnic Affairs are specifically charged with addressing black concerns. Under its terms, Afro-Colombians and other minority groups have been given special access to media. National research institutes, such as the Colombian Institute of Anthropology and History, are funding research on Afro-Colombian themes. Law 70 also spurred legislation to include Afro-Colombian history and culture in the national curriculum and to award special grants for black students to attend the university. Wade cites evidence to show that such legislation has been enacted in several cases.

Law 70 is also being used with variable success to assert black rights in other regions of the country, such as the Andean interior and the Atlantic coast. In 2005, I attended a meeting between PCN activists and Lavinia Fiori, a journalist and longtime supporter of black rights. Fiori had been working as a consultant with the Colombian National Park Service and had learned about the problems faced by black populations living in the Caribbean archipelagos around Cartagena. Recently, many of these communities were being forcibly evacuated to develop tourist resorts on the islands. Fiori wanted to strategize with Rosero about the possibility of the communities' drawing on Law 70 to resist forced evacuation. When I returned in 2007, I heard that such procedures were under way.

Black representatives are part of several government ministries (Rural Development, Environment, Education, etc.) and regional development corporations. However, many such gains are not without contradictions, as the issue of land appropriations makes evident. And as activists note, Afro-Colombians who are elected to official posts or appointed to government posts usually do not represent the poor black base of either rural or urban areas. This charge has been leveled especially severely at two recent Uribe appointees: Minister of Culture Paula Moreno, the first black woman to hold the office, and Deputy Minister of Labor Relations Andrés Palacio. In our August 2007 conversation, Rosero suggested that President Uribe had appointed these (young, middle-class) Afro-Colombians to high offices to substantiate his claim that there is no discrimination in Colombia. He saw Uribe's actions as an attempt to appease the U.S. Congressional Black Caucus as well as Afro-Colombians. A driving interest for Uribe may be the support of both for the U.S.–Colombia free-trade agreement.

Rosero acknowledged that Moreno seemed to be an intelligent, capable woman who represented a certain sector of Afro-Colombians.

However, for Rosero she was not part of the black struggle and thus did not represent "the black people." The new minister of culture was trying to "find her feet," he conceded, and to learn about the complexity of black issues in the country, but "it is not enough for the black struggle to have visibility through public figures. These black faces need to have an anti-racist politics, a position that can help the structural inequality and discrimination that our people face, that helps protect our territories and natural resources, and that allows for a real, material equality, not merely formal equality which is what these appointments are achieving." Rosero asserted it was becoming increasingly important to talk about racism and discrimination against Afro-descendants in Colombia. He continued that the PCN and the black struggle were "ten years behind" in discussing the importance of these problems. Although the PCN had entered the struggle for black ethnic and territorial rights at a different point than the Movimiento Nacional Cimarrón, which focused specifically on racial equality and discrimination, it needed to discuss these matters now. He argued that it was especially important to address the economic elements of black discrimination, something that neither Cimarrón nor the PCN had focused on in the past. This focus was the basis of the PCN's renewed collaboration with Cimarrón and its leader, Juan de Dios Mosquera.

Rosero told me that increasing numbers of young people were affiliating with the PCN and that it was important to foster their leadership skills. He also spoke of working with Afro-Colombian university students. To this end, the PCN had established an "Observatory of Racial Discrimination" in collaboration with Los Andes University in Bogotá. The observatory is a space for research and discussion to document racism in Colombia and to work against it.[17]

I asked where the issue of black reparations fit into this discussion. Rosero answered that issue of reparations in general emerged from the Peace and Justice Law (Law 975 of 2005), governing the demobilization of illegal armed groups (especially paramilitary groups).[18] The issue of reparations for black communities arose after President Uribe denied the presence of racism in Colombia. Rosero noted that it was a new issue and that its parameters and relations to other elements of the black struggle had not been worked out.

Rosero brought up an issue that has been on the PCN's agenda since the early 1990s: the need to make alliances with the African diaspora, especially in the Caribbean, to work for social and economic justice. He mentioned that through their WOLA connections, black activists had begun working with a black diaspora advocacy group called the TransAfrica Forum.[19] Rosero told me that the PCN was considering adding "connections with the diaspora" as the fifth of its focal points for action (the first four being territory, identity, development, and participation).

Speaking of these early PCN principles, Rosero said that the massive forced migration of riverine communities, the increasing complexity of black struggles and the displacement of many leaders from the coast had put downward pressure on the PCN's work at the grassroots. PCN activists recognized this problem and were working aggressively to solve it. Indeed, that very weekend Rosero and Hernan Cortés planned to visit communities in coastal Cauca and Nariño to revive their connections there. He added that, while organizations such as AFRODES worked with displaced communities, the PCN wished to concentrate on preventing displacement. But Rosero was well aware that preventing displacement and organizing grassroots black communities would be complicated tasks, with few guarantees and many risks.

Such risks largely prevented my return to the riverine communities and regional capitals in the Pacific I had last visited in 1999. My accounts of the most recent changes in organizing dynamics among grassroots communities are based on secondary sources. Anecdotes and conversations with black activists and Colombian officials, like the eyewitness accounts of church groups, journalists, human-rights observers, solidarity groups, and others visiting the field, indicate that the situation of grassroots communities in the Pacific is bleak. Yet interspersed with accounts of displacement, violence, hunger, and disenfranchisement are reports that laud the work and achievements of "alternative development projects," community councils, and internally displaced persons who have returned to establish "communities of self-determination, life, and dignity."

In the mid-to-late 1990s, community councils proliferated, as local communities mobilized to petition for collective land titles. Depending on the location, these local councils had varying degrees of contact and support from regional organizations such as the PCN and the ACIA. As violence proliferated and the regional organizations weakened and split, local councils formed without connections to regional movements. Although such independence allows for the development of leadership at the local level, I heard accounts of many councils being formed or manipulated by the army and paralegal forces. In any case, local councils are rarely able to assert their authority, either because of the displacement of leaders or because of external threats and structural factors. This meant that it was difficult for communities to organize to petition for collective titles—or to exercise any real control over their land, when and if such titles were received. Moreno, Mejía, and Rosero were among those who said that collective land titles were at best increasing "in name only."

In the southern Pacific, in Nariño and the surrounding areas, massive displacement events continue, despite the supposed demobilization of paramilitary groups. The major culprits are aerial fumigation and the subsequent (and often forcible) takeover of lands for drug cultivation or agro-industrial development. Of course, Afro-Colombian communities and organizations are not entirely passive in light of these grim events. Moreno told how the community councils in the Tumaco area had come together to form a new organization called Recompensa to propose and implement alternative development projects. Alberto Gaona spoke of the work of the former Fundación Habla/Scribe member Jaime Rivas, who has established a group called Renacientes in Tumaco.[20] The group fosters leadership through a focus on arts and cultural activities. Felipe García of the Equilibrium Foundation spoke in glowing terms about one of his organization's projects—on leadership building and productive activities—with women in Tumaco. The Equilibrium Foundation works with *piangueras* (women who collect shellfish) to help them reach the mangroves where they collect shellfish and transport their harvest to market. According to García, this project has seen phenomenal success and drawn many men into its orbit, as well. From what García told me, the project does not have any connections with the women's cooperative in Tumaco. He also said that the Equilibrium Foundation had tried to work with women's groups in other areas but had not had much success. He attributed this lack of success to resistance from menfolk and the fact that many of the women's groups were no longer focusing on productive activities. Gaona spoke in similar terms about the women's groups in Guapi. According to what he had heard, CoopMujeres was doing well, but many of the women associated with Matamba y Guasá were split over how to proceed in their activism.

Email communications from the PCN indicate that the community councils in the coastal Valle del Cauca river basins remain active in a variety of ways: attempting to keep drug cultivation and armed forces from their communities; denouncing the actions of armed forces and other forms of violence; and engaging in productive activities, including some through the kind of women's projects that Congolly was involved in. Similar activities are occurring in the Chocó. In addition to international organizations, longtime supporters of black struggles such as Jeannette Rojas continue to stand with these communities, organizing marches and other events to keep their plight in the spotlight and to remain in solidarity with them.

Some grassroots communities, such as those involved in the Patía River basin project discussed in chapter 2, also seem to have succeeded in obtaining support from state officials. During our August 2007 conversation, Mejía told me that the territorial zoning and collective titling project that had begun in the mid-1990s by the community councils of the Patía River (Black Community Council of the Greater Patia River Basin, ACAPA) was continuing with the support of Social Action's Land Protection Project. While she was unsure of the long- or even medium-term future of the project and of the fate of the Patía communities, she was happy to end our conversation on a positive note. As a parting gift, she gave me a copy of a publication entitled, "Gente de esteros, ríos y mar: Zonificación para la protección del territorio colectivo de ACAPA (People of the Estuaries, Rivers, and Ocean: Zoning for the Protection of the Collective Territory of ACAPA)."

I end this book on a positive note, with some contingently good news about Doña Naty and the other residents of Calle Larga. At the time of this writing, news reached me of an Anchicagueño among the coastal citizens gathered near Buenaventura to consult with the PCN and other activists. In November 2008, on a brief visit to the PCN offices I saw a picture of this gathering pinned on their bulletin board. Among the people in the picture was Doña Naty with a broad, brave smile on her face. A day later, a young agronomy student from Buenaventura told me that Doña Naty and Anchicagueños were still fighting for a life of dignity, peace and freedom in the rivers.

The grim reality of black communities in the Pacific lowlands of Colombia is but one illustration of how the drive toward capitalist modernity results in displacement and violent dispossession (Comaroff and Comaroff 2001; Escobar 2003). At the same time, this drive toward economic development even in its current globalizing guise reinforces state presence. It is almost a truism to say that the unfolding of such power is inevitably met with resistance. That is, "local communities" are not unproblematically and completely subsumed by the violent interventions of armed capital. Contemporary struggles for social change not only try to put a brake on the negative forces of capitalist globalization and state power; they also seek alternatives to them. But as I have argued through the case of black struggles in the Pacific lowlands, such alternatives are not simply outside or opposed to modernity; rather, the alternatives emerge from the relations between struggles and development. Understanding these relations, and the tensions and power that underlie them, requires tracing the complex, contradictory, and contingent ways in which development and resistance structure each other.

APPENDIX A

Transitory Article 55

President of the Republic of Colombia. The Political Constitution
of Colombia 1991. Bogotá, Colombia.

TRANSITORY ARTICLE 55
Within two years of the approval of the new Constitution, the Congress will
elaborate and approve, based on a study to be conducted by a special commission to be created by the Government for such purpose, a law that recognizes the
collective property rights of black communities that have inhabited the empty
lands (*tierras baldías*) in the rural riparian zones of the Pacific coast, in accordance with their traditional production practices. The area will be demarcated
by the same law.

Representatives of the black communities thus recognized will participate in
the special commission mentioned above.

The property recognized by this law is transferable only according to the
terms indicated by the law.

The same law will establish mechanisms for the protection of the cultural
identity and the rights of these communities, and to promote their economic
and social development.

PARAGRAPH I
The terms of this article can apply to other zones of the country that are determined to have similar conditions based on the same procedure and prospective
studies and favorable consent of the special commission foreseen here.

PARAGRAPH 2
If at the expiration of the term indicated by this article the Congress has not
promulgated the law to which it refers, the Government will do so within six
months through norm with the force of law.

APPENDIX B

Law 70 of 1993: Outline and Salient Features

THEME	CONTENT
Chapter 1	
Objectives (Beneficiaries)	—Recognize the black communities of Colombia as an ethnic group. —Establish mechanisms to protect their cultural identity and promote their social and economic development. —Recognize their rights to collective ownership of the *tierras baldías*, or vacant lands in the rural, riparian zones of the Pacific. This right to collective property also applies to other black communities that live in similar conditions and manage their lands using traditional practices of production (Articles 1, 2, 4–18).
Definitions	The following terms are defined in Article 2: —Pacific basin —Rivers of the Pacific basin —Rural riparian zones —*Tierras baldías*, or vacant lands —Black community —Collective occupation —Traditional practices of production

THEME	CONTENT

Chapter 2

Principles

Article defines the principles on which Law 70 is based:
—Recognition and protection of the ethnic and cultural diversity and the right to equality of all the cultures of the nation.
—Respect for the integrity and dignity of the cultural life of the black communities.
—Participation of the communities and their organizations, without detriment to their autonomy, in all the decisions that affect them.
—Environmental protection, focusing on the relations between the black communities and nature.

Chapter 3

Recognition of Collective Property Rights

Under Article 4, the land owned collectively by black communities will be denominated "Lands of the Black Communities."
Articles 5 and 46 discuss the conformation of community councils, or *consejos comunitarios*, of the black communities as forms of internal administration.
Article 6 defines the limitations and exclusions of collective property rights (public lands, urban areas, renewable and nonrenewable natural resources, indigenous resguardos, subsurface resources, areas denominated as security or national-defense zones, national parks).
Articles 6, 14, and 19 define the ecological functions of the collective property.
Article 7 determines that the Lands of the Black Communities will be legally inalienable (*imprescriptibles, inembargables*).
Articles 8–13 and 16 describe the procedures to obtain collective property rights with the help of INCORA, IGAC, and INDERENA (later the Ministry of the Environment).
Articles 15, 17, and 18 describe the mechanisms by which the collective property rights will be protected.

THEME	CONTENT

Chapter 4

Land Use and the Protection of Natural Resources and the Environment

Article 19 sanctions traditional practices of land use for subsistence.

Articles 20–25, 29, and 49 denote the social and ecological functions of collective property rights.

Articles 22, 23, and 25 outline mechanisms to deal with the overlap between national parks and areas occupied by black communities.

Article 24 outlines the mechanisms by which the extraction of forest products, especially timber, from the collective land is to be approved by the entity in charge of regulating the processing and marketing of such products.

Special natural-reserve areas are to be defined jointly (Articles 25 and 51).

Chapter 5

Mineral Resources

Article 27 delimits the mining zones of black communities and discusses the special permits that determine the rights to explore and exploit mineral resources in the Lands of the Black Communities.

Under Article 28, indigenous and black mining lands can be declared "Joint Mining Zones."

Article 29 describes the limits to mining uses.

Article 30 sets the control mechanism over mining contracts.

Article 31 discusses the implementation of the mining rights of black communities.

THEME	CONTENT

Chapter 6

Mechanisms to Protect Cultural Identity

Articles 32, 34–36, and 38–39 focus on the ethno-education rights and discuss mechanisms to develop ethno-education strategies.

Article 33 describes the sanctions against discriminatory practices.

Article 37 discusses the various ways in which information regarding Law 70 will be disseminated among the black communities.

Articles 38 and 40 discuss access to technical, technological, and professional training and the creation of a special scholarship fund.

Article 41 determines how organizational processes are to receive assistance.

Article 42 decrees the formulation and execution of an ethno-education policy and the formation of a Pedagogy Advisory Commission.

Article 43 restructures ICAN to include an Afro-Colombian Research Program.

Article 44 decrees the participation of black communities in environmental, socioeconomic, and cultural-impact studies.

Articles 45 and 60 establish the High-Level Consultative Commission to coordinate and recommend ways to implement Law 70.

Article 62 decrees that the government put forward funds within two years to inaugurate the University of the Pacific that was created under Law 65 of December 14, 1988.

THEME	CONTENT

Chapter 7

Planning and Promotion of Social and Economic Development

Under Article 47, the black communities have rights to development according to their culture.

Articles 48 and 56 determine participation in the National Planning Council, the Territorial Planning Councils, and the executive councils of regional-development corporations.

Article 49 determines the right to participate in the discussions of development plans, programs, and projects.

Articles 50–51 promote participation and research.

Article 52 encourages the creation of production associations, including funding cooperatives.

Article 54 legally recognizes the traditional knowledge of black communities and assures them the economic returns from any product developed by them that is patented.

Article 55 establishes credit and technical-assistance programs.

Article 57 decrees the creation of a study commission to formulate a Development Plan for Black Communities.

Article 58 creates project-management units in the state social investment funds.

Article 59 determines that hydrographic basins can be considered planning units.

Chapter 8

Other issues

Articles 6 and 64 allocate resources for the execution of Law 70.

Article 66 reserves two seats in the Chamber of Representatives for black candidates.

Article 67 creates a special director of black community affairs in the Ministry of Government.

NOTES

Introduction

1. See appendix A for the full text of AT 55.

2. The Chocó region of Colombia includes the Cordillera Occidental; the Serranías of Baudó and Darien; the Atrato, San Juan, and Baudó river basins; and the upper Sinú and San Jorge regions of western Colombia. The departments of Antioquia, Córdoba, Risaralda, Chocó, Valle del Cauca, Cauca, and Nariño are within the political and administrative limits of the region labeled the Colombian Pacific. The 8 million hectares of the Colombian Chocó (6.2 percent of the country) encompass 5.5 million hectares of forests, of which 3.5 million are considered pristine or without major interventions. This natural-resource-rich area supplies 60 percent of the total timber demands of Colombia, 82 percent of its platinum, 18 percent of its gold, and 14 percent of its silver (Tulio Díaz 1993). In 1993, 2.5 million hectares of this area were in eight national parks, and 1 million hectares were in seventeen indigenous resguardos, or indigenous communal lands. The rest was considered *tierras baldías*, or empty lands, belonging to the state, even though they were inhabited by a significant percentage of Afro-Colombian communities.

3. See appendix B for a list of the salient features of Law 70.

4. Beyond these binaries, the 1990s also generated many critical debates and insightful accounts of the histories and contingent relations between development and modernity (Cooper and Packard 1997; Cowen and Shenton 1996; Crush 1995; Edelman and Haugerud 2005; Leys 1996).

5. These included Nicaragua (1987), Argentina and Bolivia (1994), Brazil and Ecuador (1988), and Venezuela (1999).

6. The absence of military authoritarianism and the presence of formal democracy notwithstanding, these problems and protests took multiple violent forms. See Bergquist 2001; LeGrand 2003; and Palacios 2006 for assessments of the complex factors shaping violence in Colombia and how historical factors bear on this conflict in the twenty-first century.

7. From the end of the nineteenth century to the mid-twentieth century, coffee production and export, and the revenues from the same, played an important role in Colombian economic history and development. For more on the role of coffee in the Colombian economy, see Jiménez 1995; Palacios 1980. On the relationship between coffee production and violent land struggles, see also Bergquist 1986.

8. For further details on the conditions under which the guerrilla groups agreed to lay down their arms and participate in Colombian democratic politics, see Arocha 1992: fn. 1, 1994: 94. For the development of M-19 prior to the peace accord of the 1980s, see Fals Borda 1992. Among other sources, the last chapter of Bushnell 1993: 252–59 and Bergquist et al. 2001 offer discussions of the role of left-wing guerrilla groups in modern Colombian politics.

9. See Agudelo 2001, 2005; Dugas 1997; Palacios 2006; and Van Cott 2000 for detailed discussions of the constitutional-reform process and of the origins of the 1991 Constitution. According to Palacios (2006), the lead-up to the election of the Constituent Assembly was dubiously legitimate, and the Constituent Assembly was elected with the lowest turnout (26 percent) in Colombian history. He further notes that the assembly overstepped its mandate: rather than modifying the old constitution, it drafted a new one. The resulting constitution—a mosaic of 380 permanent articles and 60 provisional articles, was sloppily and unevenly drafted. See also Bejarano 2001 for an assessment of the new constitution and why almost a decade later it was failing to meet its framers' expectations. Palacios (2006) also addresses this issue.

10. I use the term "indian" because anthropological and postcolonial work in the past decade has shown that "Indian" identity is as multiple and mobile as other cultural categories such as "black" and "mestizo."

11. Beyond the small body of classical anthropological studies on black communities in Colombia (de Friedemann 1985a, 1985b, 1989; de Friedemann and Arocha 1984, 1986, 1995; Whitten 1986), an important body of literature on Afro-Colombians has emerged since the 1990s. It offers rich and extensive accounts of how black groups in Colombia are (re)constructing their sense of ethnic identity, tradition, and struggle by mobilizing memory; ideas of culture, community, and history; and connections to place, space, and nature (Hoffman 1999, 2002; Losonzcy 1991, 1993, 1999; Oslender 1999, 2001; Restrepo 1996, 1998, 1999, 2001, 2004; Villa 1998, 2001; Wade 1993a, 1996, 1999, 2002).

12. My reading of Afro-Colombian struggles as an example of "cultural politics" is shaped strongly by feminist theorists (Ferguson 1993; Flax 1990; Scott 1988, 1990; Westwood and Radcliffe 1993) and cultural-studies approaches (Anthias and Yuval-Davis 1993; Bhabha 1994; García Canclini 1989; Gilroy 1987; Grossberg et al. 1992; Hall 1986, 1992a, 1992b; Rutherford 1990).

13. Arturo Escobar's observations and analyses have generated much debate within political ecology, which is not my central concern here: see Escobar 1999 and the comments and exchanges on it in *Current Anthropology* 40(1): 1–30.

14. My theoretical understanding of these linkages is informed by many beyond those cited in the text, including those who (following Marx's maxim

about the ruthless criticism of everything) offer deep and extensive insights into the processes and crises of capitalist modernity and development (e.g., Harvey 1989; Laclau and Mouffe 1985; Roseberry 1989; Spivak 1999).

15. The feminist political theorist Jane Flax notes that, although there is no united or coherent approach that can be labeled "postmodernism," in general postmodernists "reject representational and objective or rational concepts of knowledge and truth; grand, synthetics theorizing meant to comprehend Reality as and in a unified whole; and any concept of self or subjectivity in which it is not understood as produced as an effect of discursive practices" (Flax 1990: 188).

16. The parameters of postcolonial theories and approaches are subject to much debate. Escobar cites the influences of the writings of Edward Said, V. Y. Mudimbe, Chandra Mohanty, Timothy Mitchell, and Homi Bhabha on his work. See Gupta 1998: 6–8 for an overview of some of the questions about the term "postcolonial" and for a list of references that address these questions.

17. Exploring the dangers of post-developmentalist discourses of "radical," non-Western difference, Meera Nanda (2002) asks how such discourses play out, if subalterns actually embrace them, and whether recognizing "local" difference redistributes economic and cultural power.

18. My debt to feminist scholarship is as wide as it is deep, as it has explicitly or implicitly informed all of my work. Among the works that I have returned to repeatedly are Agarwal 1992; Anthias and Yuval Davis 1993; Haraway 1991; Mohanty 1997, 2003; Mohanty et al. 1991; Mufti and Shohat 1997; Scott 1988, 1990, 1992; Spivak 1989, 1990.

19. See, e.g., Alvarez 1990, 1998; de la Cadena 2000; Lind 2005; Schild 1998; Thayer 2001; Westwood and Radcliffe 1993.

20. See Wainwright 2008 for an analytical engagement with development's aporias.

Chapter 1: Afro-Colombian Ethnicity

1. For some works related to Latin America and including Colombia, see Appelbaum 2003; Applebaum et al. 2003; Grandin 2000; Leys Stepan 1991; Sanders 2004.

2. In one of the first comprehensive histories of Afro-Latin Americans, George Reid Andrews (2004) argues that black groups in Latin America are more likely to be involved in broader democratic politics in the region than to engage in racially oriented movements.

3. Indigenous people in Colombia have a fair degree of autonomy (at least on paper) to manage their lands and administer their territories. For example, in the early 1990s about three hundred resguardos or indigenous territories of various sizes were recognized in Colombia. Some are as large as 5 million hectares, as in the Putumayo region of the Amazon; others are very small areas of indigenous lands surrounded by the *campesino fincas* in the north of Cauca. In all, about 20–24 percent of the country—that is, between 25 million and 28

million hectares—were under the jurisdiction of various indigenous groups, who form about 2 percent of the Colombian population (Roldán 1993; Sánchez et al. 1993).

4. Demographic data on what percentage of the country or of the Pacific region is "black" is highly unreliable and varies significantly. The census conducted in 1991 was declared so full of errors that the Colombian state reformulated its census forms and collected new census statistics. According to national data collected in 1993, the Chocó biogeographic region has a population of 2.3 million (a little more than 6 percent of the total population of Colombia), of which 90 percent are Afro-Colombian and 4 percent are indigenous. It estimates that 1.3 million of these 2.3 million people live in the rural areas of the Pacific and represent 12 percent of the total rural population of the country. The 2005 census indicates that 10.5 percent of Colombia's population, just under 4.3 million people, is Afro-Colombian.

A number of other figures prevail among scholars writing on Afro-Colombians. Motta (1992) estimates that 4 percent of the total Colombian population, or 5 million of Colombia's 36 million inhabitants, are black. Lozano (1992) estimates the total black population at 21 percent; Restrepo (1992) estimates it at 24 percent; and Perea (1990) estimates it at 30 percent. None of these authors cite their sources. The Colombian National Department of Statistics (Departamento Administrativo Nacional de Estadística), the agency responsible for collecting national census and demographic data, has a number of different papers discussing the problem of accounting for "blacks" in the country: see the website at www .dane.gov.co. See Urrea et al. 2001 for a discussion of the problems of collecting demographic information on blacks in Colombia, and see Wade 2002: fn. 6 for other comparative figures.

5. Vargas (1993) discusses at length the territorial dynamics in the Río Atrato basin of the Pacific region during the time of the Spanish invasion and its impact on the native population of the region. Colmenares (1982), de Granda (1988), Hudson (1964), and Jaramillo Uribe (1963) give detailed historical analyses of various aspects of the slave economy, social relations during slavery, and the colonial economy. See also Jaramillo Uribe 1989, which examines and summarizes the earliest historical works on Afro-Americans and Afro-Colombians. See Aprile-Gniset 1993; Romero 1990–91, 1995; and Zuluaga 1994 for detailed discussions of the population and resettlement of the Pacific region by black groups.

6. *Paros cívicos*, or civil strikes, occurred all over the region from the 1980s to well after the passing of Law 70. Murillo 1994 argues these strikes are part of a tradition of black social protests against state neglect and lack of basic social and public services that have occurred from the northern Chocó to southern Tumaco since the 1950s.

7. The journal *Esteros*, funded by the Swiss development agency SWISSAID, published a number of articles on the problem with logging concessions in the region: see Gómez 1994; López 1996; Ríos 1993a, 1993b, 1994.

8. Multiple conversations with Alberto Gaona in Cali in 1993, 1994, and 1995. According to Gaona, among Fundación Habla/Scribe's founding members were

progressive academics from the Universidad del Valle and professionals who had worked among black and indigenous communities in or from the Pacific region. These included Gustavo de Roux, Alvaro Pedrosa, and María Teresa Findji.

9. In addition to two indigenous candidates, this seventy-member body included an evangelical pastor; more than thirty ex-guerrillas from the M-19 alliance, the Popular Liberation Army, and the Revolutionary Workers Party; and an indigenous representative from the armed Indianist group Manuel Quintín Lame (Arocha 1992).

10. Arocha (1994) discusses how the indigenous delegates and the sociologist Orlando Fals Borda of the Constituent Assembly were invited to a meeting entitled "Blacks in the Constitution" organized by the group Viva la Ciudadanía in Cali in May 1991. Arocha notes bitterly that the eagerly awaited Constituent Assembly delegates never showed up; nor did they make a case for the ethnic rights of black groups in their proposals to the Subcommission on Equality and Ethnic Rights. Wade (1995) argues that this kind of lack of support for the ethnic claims of black communities indicates that the strength of ethnic alliance between Indians and blacks was overstated. Grueso et al. (1998) notes that, although the progressive sectors of the Constituent Assembly, including members of leftist groups, were sympathetic to the socioeconomic and political plight of blacks, they were unwilling to recognize the cultural difference and ethnic specificities of Afro-Colombians.

11. Texts of the documents prepared by Fals Borda and Muelas and presented to the subcommission were later compiled by the anthropologist Myriam Jimeno and appeared in *America Negra* 3 (1992).

12. See Cano Correa and Cano Busquets 1989 for coverage of a discussion among some of the members of the Constituent Assembly, including leading Colombian academics and social activists, that followed the publication of Arocha 1989.

13. Among the twenty-eight members of the CECN were twelve black representatives, two each from the departments of Atlantico, Antioquia, Cauca, Chocó, Nariño, and Valle del Cauca; two scholars on Afro-Colombian issues; and representatives of the state's Ministry of Government, IGAC, INCORA, INDERENA, ICAN, and DNP. In addition, certain members of the Chamber of Representatives and the Senate, as well as certain politicians from the Pacific region, participated in CECN discussions. See Miguel Vásquez 1994: 51–52 for the text of the decree and further details on the members of the CECN.

14. Although much has been written about this period of black mobilization, authorship of these drafts and the final law are not attributed to a single person or entity. On January 14, 1995, I spoke to Miguel Vásquez, an indigenous rights lawyer who had worked with the Awa in Nariño and who served as a legal adviser to Zulia Mena. According to Vásquez, he drafted Law 70 with some key leaders of the OCN and members of the CECN who were in solidarity with black struggles for ethnic and territorial rights.

15. For a discussion of the situation of blacks in the Colombian-owned Caribbean islands of San Andres and Providencia, see Gallardo 1986; Pedraza Gómez

1986; and Villa 1994. See also Ng'weno 2007 for a discussion of the complexities of applying Law 70 among black communities in the agro-industrial sugar zones and mining areas of the northern Cauca Valley, where the traditional norms of land use and property rights differ from those of the Pacific region.

16. See Velasco and Preciado 1994 for details of the electoral campaign of these and other black candidates during the 1994 parliamentary elections. Velasco and Preciado also discuss the effect of ethnic and minority candidates in recent Colombian electoral politics.

17. In 1998, Afro-Colombia lost the special circumscription seats on a technicality. However, when the special circumscription was reinstated in the 2002 election, María Isabel Urrutia and Willington Ortiz, two black athletes, won the seats. Neither had any prior experience with black struggles; nor had they emphasized any pro Afro-Colombian rights position during their campaigns. By the 2006 elections, the PCN was reconsidering its strategy, and Carlos Rosero ran (unsuccessfully) for elections.

Chapter 2: "The El Dorado of Modern Times"

1. These descriptors are prevalent in state reports, national newspapers, and television news and popular programs in the Chocó region. I heard similar descriptors and sentiments expressed by local people especially in Bogotá, but also in Cali. Arturo Escobar and Alvaro Pedrosa (1993: 44, fn. 2) note that these terms originated at the third forum, "Colombia in the Era of the Pacific," held in Popayán on May 13–15, 1993.

2. The following are some of the dismal development indices for the Pacific. Only 47.2 percent of Chocoans have access to potable water. The Pacific coast has the second-highest mortality rate in the country for children under five, at 32 per 1,000 live births. Chocoans have the lowest life expectancy (age 66.6) of all Colombians.

3. According to Escobar (1995), a constitutional reform in 1945 introduced the concept of "planning," and since the 1950s, each Colombian administration has formulated a development plan for the country. In 1966, the government established the National Council for Economic and Social Policy (Consejo Nacional de Política Económica y Social, CONPES) to aid the DNP in this process.

4. Autonomous regional development corporations are private entities with a public character designed along the lines of the Tennessee Valley Authority in the United States. In addition to the CVC, other corporations involved in Pacific development include CORPURABA, CODECHOCO, the CRC (for Cauca), and CORPONARIÑO.

5. For an excellent overview of international development efforts in the Pacific region, see Carpay 2004.

6. The GEF is a funding program from industrialized nations established in the early 1990s under which grants and loans are provided to developing countries to help them carry out the tasks of environmental protection. The GEF is

administered by the World Bank, with collaboration from the United Nations Environmental Programme and the United Nations Development Programme (UNDP). About 30–40 percent of the total GEF funds are allocated for biodiversity conservation.

7. In addition to the Convention on Biological Diversity, four other international accords were discussed at ratified at UNCED, held in Rio de Janeiro in 1992. See Haas et al. 1992 for a summary of the principal accords, including the Convention on Biological Diversity, discussed and ratified at UNCED.

8. The other research institutions are the Instituto de Hidrología, Meteorologia y Estudios Ambientales, which will be responsible for research on hydrology and meteorology; the Instituto de Investigaciones Marinas y Costeras "José Benito Vives de Andries," which will focus on marine and coastal research; the Instituto de Investigación de Recursos Biológicos "Alexander von Humboldt," which will coordinate all biodiversity research in the country; and the Instituto Amazónico de Investigaciones Científicas "Sinchi," which will coordinate all research activities in the Colombian Amazon region.

9. Members of the evaluation team present at the January 28, 1995, meeting were Peter Wilshusen, an environmental scientist from the University of Michigan, and Julio César Tresierra, a Peruvian Canadian sociologist from Concordia University in Montréal who was an independent consultant with the UNDP; Edgar Cortés from the DNP; Jorge "Mono" Hernández, a well-known Colombian biologist who was a consultant to the government; and Alan Bidaux, a representative of the Swiss government.

10. "El Ordenamiento Territorial es una política de estado y un Instrumento de planificación, que permite una apropiada organización político- administrativa de la Nación, y la proyección espacial de las políticas de desarrollo social, económico, ambiental y cultural de la sociedad, garantizando un nivel de vida adecuado para la población y la conservación del ambiente" (IGAC 2001, definición, para. 1).

11. At IGAC, I also spoke to Manuel Amaya Amaya and Alberto Boada on May 25, 1995. In August 1995, I heard Cesar Monje and Ricardo Castillo present and discuss their pilot study to measure migration and resettlement in the Atrato region of the Chocó. Much of what I heard from them was reiterated in IGAC publications. I cite them where relevant.

12. Many of these were technical geographic information system projects undertaken to outline Colombia's PAFC under the PMRN. Details of the projects and project results are in *Revista Informativa del Proyecto SIG-PAFC*," published periodically between 1994 and 1997.

13. For the results of some of these studies, see Arroyo and Aguilar Ararat 1995; Camacho and Tapia 1996; Olander 1995; Sánchez 1995.

14. In 1995, I also had several conversations about the project with Clara Llano in Cali; with Miguel Vásquez (on January 14); and with Eduardo Ariza (on April 27). Llano also shared the final project report with me.

15. Llano's fiancé was a television producer and had helped get Mena's campaign advertisements on TV.

16. Cf. Fiori 1995; Hurtado 1996; Pacheco and Achito 1993; Pacheco and Velásquez 1993; Pardo 1996, among others, for discussion of inter-ethnic relations and alliances between indigenous and black communities in the Pacific.

17. See Castro Hinestroza 1996 for a detailed comparison of these draft statutes, as well as the draft produced by some NGOs. Castro Hinestroza also discusses the international and national environmental accords that shaped the establishment of the IIAP.

18. "Son fines esenciales del Estado: servir a la comunidad, promover la prosperidad general y garantizar la efectividad de los principios, derechos y deberes consagrados en la Constitutión; facilitar la participación de todos en las decisiones que los afectan y en la vida económica, política, administrativa y cultural de la Nación; defender la independencia nacional, mantener la integridad territorial y asegurar la convivencia pacífica y la vigencia de un orden justo. . . . Las autoridades de la República están instituidas para proteger a todas las personas residentes en Colombia, en su vida, honra, bienes, creencias, y demás derechos y libertades, y para asegurar el cumplimiento de los deberes sociales del Estado y de los particulares" (Constitución Política de Colombia 1991, Artículo 2).

Chapter 3: "El Ruido Interno de Comunidades Negras"

1. In Gramscian terms, the PCN seemed to want to engage in the "war of positions" and the "war of maneuver" simultaneously. PCN leaders knew that the former was a protracted struggle across different fronts and that Law 70 was not a definitive victory in the "war of maneuver."

2. See Wade 1999 for a longer discussion of Grupo Ashanty and other music and dance troupes from Cali. Wade argues that, despite their claims, they are celebrating black identity and making a space for black culture in the urban context, and their politics is as much a reflection of class as it is of black culture.

3. *Palenques*, or colonies of escaped slaves, were found throughout the Americas. They were called *palenques* in Mexico and Cuba; *cumbes* in Venezuela; and *quilombos*, *mocambos*, *ladeiras*, and *mambises* in Brazil. In the Caribbean, the Guyanas, and the U.S. South, they were known as maroons. See de Friedemann and Cross 1979 for a detailed study of the palenques in Colombia.

In the southwestern Pacific, the Palenque de Castigo was one of the best-known refuges in Río Patia for slaves who escaped from the mines. According to Zuluaga (1986), it was established around 1732 at the same time that mining activities increased in the Patía region. See also Zuluaga 1994 for a discussion of the formation and consolidation of communities of freed or escaped blacks in the Pacific during slavery. According to Cortés (1994), the palenques in the northern Pacific (like the famous Palenque de San Basilio near Cartagena) were primarily fortified camps of organized resistance to slave holders, while the southwestern palenques were primarily settlements housing free blacks.

4. In 1995, I conducted fieldwork in several black communities in the Pacific region and had follow-up conversations with members of research teams who

continued to work at these sites. I visited two communities in Río Satinga and Saquinga in Nariño in March 1995 with agronomists from the Guandal Project. I spent ten days in three communities in coastal Chocó in June 1995 in collaboration with Fundación Natura, conducting natural-resource-use workshops and talking at length with women in the communities. I made several stays in Calle Larga from June to October 1995, and my work there was enhanced by conversations with a research team of agronomists, agricultural economists, and foresters and a social scientist from Fundación Herencia Verde who had been working in the region for several years. Conversations with Eduardo Restrepo, Juana Camacho, and Jorge Ceballos have been invaluable in outlining issues of land and resource use and property rights among Pacific black communities.

5. Rhymes, poems, stories, jokes, and gossip form integral parts of Afro-Colombian oral culture, which is still vibrant in the region. Coplas are couplets or popular songs mostly composed and sung by women, and a person proficient in the art of writing them is called a *coplera*.

6. During my fieldwork in Calle Large, Doña Naty took me to visit her blind mother-in-law. She introduced me as "La Kiran, una blanca pero buena (Kiran, who is a white but good person)." When I asked Doña Naty why she had introduced me this way, she said that she could not call me a *paisa* (a person from Antioquia), because the *paisas* in Anchicayá were merchants and miners and were mistrusted by the riverine blacks. She could not call me an *india* or *chola* because, like many blacks, she was ambiguous about the social status of indigenous people. She did not want to call me a *negra* because my skin was much lighter than hers. All of those things she could not call me because she did not want to insult me, and for the same reason she could not call me "white" without a qualifier. So for the blind mother-in-law, I became "*blanca pero buena*."

Chapter 4: "Seeing with the Eyes of Black Women"

1. For discussions of black women's roles in Colombian history and society, see Espinosa and de Friedemann 1993; Gutiérrez de Pineda 1994 (1968); Motta 1994; Perea 1986; Romero 1995. Camacho (2004) provides an excellent overview of the available literature on Afro-Colombian women, reviewing these and other recent sources, such as theses and unpublished talks.

2. During the United Nations' second Decade of Development (1971–80), a cadre of professional women from the non-Western world and Western-based feminists played an instrumental role in focusing attention and resources on "Growth with Equity" programs for women. The first United Nations Conference on Women was held in 1975 in Mexico City, with the declaration of 1975 as the International Women's Year. Subsequently, the years 1976–85 were declared the United Nations Decade for Women. The original WID demand for equity was reformulated at the Mexico City conference as the need to alleviate poverty among women, because governments and development agencies felt that the demand for equality between the sexes was associated with Western feminism.

The demand for equity was later linked to the argument of economic efficiency, in which women were labeled a valuable "resource" to be "harnessed" for economic development (Braidotti et al. 1994: 80). The rhetoric of women as welfare recipients and reproducers changed to women as key players and economic producers in the development process.

3. My discussion of women's cooperatives is informed by meetings with Sylveria Rodríguez, Cipriana Diuza Montaño, and other members of CoopMujeres in Guapi on April 3–4, 1995; Patricia Moreno, Dora Alonso, Mercedes Segura Rodríguez, and Myrna Rosa Rodríguez of Fundemujer in Buenaventura on August 23, 1995; and Jocelina Ceballo, Nympha Aristisaval, and other members of Ser Mujer in Tumaco on October 2, 1995.

4. Gender relations are considered a subset of the social relations of power and dominance at the household, community, regional, local, national, or international level that shape and limit women's "productive" work, as well as their "reproductive" work. For detailed discussions of how women and gender issues are addressed in development institutions, including overviews and critiques of the WID and GAD approaches, see Braidotti et al. 1994; Kabeer 1994; Moser 1993; Sen and Grown 1987; Tinker 1990. See also Asher 2004 for a discussion of how black women's activism helps contest and stretch feminist theorizing.

5. This brings to mind Audre Lorde's reflections on the black movement in the United States: "Within Black communities where racism is a living reality, differences among us often seem dangerous and suspect. The need for unity is often misnamed as a need for homogeneity, and a Black feminist vision mistaken for betrayal of our common interests as a people" (Lorde 1992: 51).

6. See Asher 1997 for a discussion of how gender metaphors are mobilized in the current black movement to define black ethnic identity.

7. Conversations with Mercedes Segura of Fundemujer and the OCN, Buenaventura, February 5, 1995; Rosemira Valencia, Cimarron, Quibdó, March 8, 1995; Victoria Torres, Women's Section, ACIA, Quibdó, March 9, 1995; Amitzury Montaño Paredes and Edith Cuero of Asojunpro Guapi, March 31, 1995; Libia Grueso, February 11, 1995.

8. See Asher 1997 for a discussion of the contradictory issue of the presence of black women as leaders in black social movements and their invisibility as black women in these organizations.

9. According to Andrews (2004: 189), black women were ignored and marginalized in other Afro-Latin American movements, as well. He notes that black women held encuentros and established separate organizations in Brazil, Venezuela, Costa Rica, Panama, and the Dominican Republic.

10. This seminar/workshop was held in Cali on January 22–24, 1996, and organized by the Program for Black Women and the Center for the Study of Women, Gender, and Society at UniValle. *Esteros* is a journal about the Colombian Pacific that was founded and has been edited by the anthropologist William Villa since 1993 and is funded by SWISSAID.

11. Norma Rodríguez, president of the Foundation for Black Ecuadorian Culture, also notes that wearing "African hairdos (*peinados africanos*)" is an original strategy to revalidate black cultural identity and build the self-esteem of black women (Sala 1995).

12. The discussion of the Black Women's Network in Guapi is based on conversations with Teófila Betancourt in August 1995 and with Yolanda García, Luz Marina Cuero, Betancourt, Eden, and other members of the network in Guapi, Cauca, on July 9, 1999. I also draw on information from published and unpublished manuscripts about and by the women's network and a videotape of the meeting in Timbaquí in May 1997.

13. See Asher 2004 for a longer, critical analysis of the involvement of black women (including those of Betancourt's group) in sustainable development and biodiversity-conservation projects in Cauca. See Camacho (1999, 2001) for a discussion of the key role of black women in managing agricultural resources and biodiversity and maintaining social networks in the coastal areas of Chocó.

Chapter 5: Displacement, Development, and Afro-Colombian Movements

1. The series of articles and maps in the special issue of the *Canadian Journal of Latin American and Caribbean Studies* on Colombia (vol. 28, no. 55–56 [2003]) are an excellent source for non-Colombianists who want to gain a better understanding of the multiple origins of the intractable, complex, and longstanding conflict in Colombia. See especially LeGrand 2003 for a historical perspective on the Colombian crisis; the ten maps in Leal 2003 visually locate illegal armed activities over time; Bejarano 2003 gives an overview of the many protaganists, manifestations, and effects of what she sees as a largely political conflict.

2. Discussions about Plan Colombia began in 1999. In July 2000, President Clinton signed legislation to give more than $1.3 billion in aid to Colombia, its neighbor, and U.S. anti-drug agencies over the next two years.

3. Many of the Colombian and non-Colombian military and ex-military contractors from the United States are graduates of the infamous former School of the Americas in Fort Benning, Georgia. In 2001, the School of the Americas was renamed the Western Hemisphere Institute for Security Cooperation.

4. The Uribe plan included a then classified operation, Plan Patriota, launched in 2003 to engage in secret dialogues with paramilitary groups to demobilize them. Like the FARC, paramilitary groups are on the U.S. list of "terrorist" organizations. However, the Colombian military's links to paramilitary groups are increasingly well documented, and both the U.S. and Colombian governments are far more lenient with paramilitary groups than with insurgent ones.

5. Citing Center for International Policy data, the COHA brief notes that in 2005, $336.1 million went to the Andean Counter-Drug Initiative; approximately $100 million went to military aid; and about $4 million went to anti-terrorism

assistance. As with the initial installment, most of this aid is given in the form of equipment and training. A mere $131.3 million went to social aid. For more on U.S. policy in Colombia and debates about U.S. aid to it, see www.ciponline .org/colombia.

6. The social and environmental damage being wrought by Plan Colombia is well documented. See Peterson 2002 for one evaluation of Colombia's war casualties.

7. While some of the armed guerrilla groups had a social mission at one time, most no longer have legitimacy or credibility, as Javier Giraldo (1999), a Jesuit priest and a long-term chronicler of the violence and crisis of justice in Colombia, persuasively argues. Through the Justice and Peace Commission of Colombia, Giraldo and others have traced the close relations between paramilitary groups and state military forces. Many observers and analysts (e.g., Delacour 2000; Richani 2002) have also shown the connections between the various armed forces, coca cultivators, and new economic interests that follow in the wake of these forces.

8. Copies of most of the National Development Plans are available at the DNP's website at http://www.dnp.gov.co.

9. As Richani (2002) notes, paramilitary forces have existed in some form in Colombia for a while. In the 1970s, their chief role was to protect the interests of large landowners; in the 1980s, new groups were formed to address the needs of the emerging class of drug traffickers. But until the 1990s, their numbers were small. Richani notes that the AUC, the most prominent of today's groups, had only ninety-three men in 1986. The paramilitary acquired a national presence and a unified command in the 1990s and rose dramatically in strength under the leadership of the infamous Carlos Castaño.

10. In inland Cauca, the Nasa indigenous communities were facing displacement as grave as that faced in coastal Pacific areas other than Cauca.

11. Like other Afro-Colombian communities and community organizations, CAVIDA was receiving support from various national and international entities. Its legal action against state entities was carried out in conjunction with the Inter-Ecclesiastic Commission for Peace and Justice in Colombia. The Cacarica communities also had links with Chicagoans for a Peaceful Colombia, who had organized a solidarity conference with community members in Chicago in April 2003.

12. According to Law 387, "Persons displaced because of violence or the threat of violence due to internal conflict, generalized violence, massive violations of human rights, or violations of International Humanitarian Law" must officially register as internally displaced persons with the *Sistema Unico de Registro* to receive government services. These services are provided through welfare agencies such as Social Action, which maintains a national database and coordinates attention to internally displaced persons through the National System for Unified Attention to the Displaced Population (Sistema Nacional de Atención Integral a la Población Desplazada).

13. According to Mejía, Social Action oversees the project and works in partnership with eighteen institutions, including international entities, state agencies such as the Office of the Attorney-General and the Office of Human Rights Protection, and civil society organizations.

14. Among the most notable are the UNHCR, Doctors without Borders, Amnesty International, Project Counseling Services (Sweden), Peace Brigades International, the American Friends Service Committee, and the Washington Office on Latin America (WOLA). A host of other organizations also maintain a presence in Colombia.

15. A variety of national and international NGOs note that the definition of internally displace people is interpreted very narrowly in Law 387, and a large number of vulnerable people are denied access to emergency assistance immediately after displacement; access to health, education, and housing services; participation in training and income-generation programs; and other forms of social support. One international NGO, Refugees International, noted that about a third of the participants at its town-hall meetings claimed to have been denied services because their names had ostensibly disappeared from the official database: see http://www.alertnet.org/thenews/fromthefield/219053/117693037949.htm.

16. Urrutia's contributions were bids for greater recognition and rights for women and athletes. Ortiz pushed for educational opportunities for blacks and for the recognition of various black cultural forms through the national recognition of the Festival of the Currulao (an important black musical form) and the feast of Saint Francis of Assisi (San Pacho), the patron saint of Quibdó.

17. For more about the aims and activities of the Observatory on Racial Discrimination, see http://odr.uniandes.edu.co.

18. For an overview of Law 975 and its shortcomings, see United Nations High Commissioner for Refugees 2006 and Amnesty International 2005.

19. For more on the TransAfrica Forum and its involvement with Afro-Colombians, see http://www.transafricaforum.org; for WOLA, see www.wola.org. Both WOLA and the TransAfrica Forum are active members of the Network for Advocacy in Solidarity with Afro-Colombian Grassroots Communities. Many other NGOs also form part of this network, which is based in Washington, D.C.

20. The term "*communidades renacientes* (emerged or reborn communities)" is gaining new currency in reference to black communities. It stresses the distinct forms of identity and cultural practices among these groups. The Colombian anthropologist Nina de Friedemann's work on black miners in Barbacoas and Eduardo Restrepo's more recent work on black communities in Nariño discuss the meaning and nuances of the term "*renaciente*" as applied to and used by local Afro-Colombian communities.

REFERENCES

Agarwal, Bina. 1992. "The Gender and Environment Debate: Lessons from India." *Feminist Studies* 18 no. 1, 119–58.

Agudelo, Carlos E. 2001. "Nuevos actors sociales y relegitimación del estado." *Análisis Político* 43 (May–August), 3–31.

———. 2005. *Retos del multiculturalismo en Colombia: Política y poblaciones negras.* Medellín: La Carreta Social Editores.

Alexander, M. Jacqui, and Chandra Talpade Mohanty, eds. 1997. *Feminist Genealogies, Colonial Legacies, Democratic Futures.* New York: Routledge.

Alto Comisionado de las Naciones Unidas para los Refugiados. 2007. "ONU pide respeto a la dignidad de afrocolombianos," 21 May. Report, on file with author. http://www.acnur.org.

Alto Comisionado para los Derechos Humanos. 2005. "Consideraciones sobre la Ley de 'Justicia y Paz,'" Bogotá, 27 June. Report, on file with author. http://www.hchr.org.co.

Álvarez, Manuela. 2000. "Capitalizando a las 'mujeres negras': La feminización del desarrollo en el Pacífico colombiano." *Antropologías transeúntes*, ed. Eduardo Restrepo and María V. Uribe, 265–287. Bogotá: ICANH.

Alvarez, Sonia E. 1990. *Engendering Democracy in Brazil: Women's Movements in Transition Politics.* Princeton, N.J.: Princeton University Press.

———. 1998. "Latin-American Feminisms 'Go Global': Trends of the 1990s and Challenges for the New Millennium." *Cultures of Politics, Politics of Culture: Re-visioning Latin American Social Movements*, ed. Sonia E. Alvarez, Evelina Dagnino, and Arturo Escobar, 293–324. Boulder, Colo.: Westview Press.

Alvarez, Sonia E., Evelina Dagnino, and Arturo Escobar. 1998. "Introduction: The Cultural and the Political in Latin American Social Movements." *Cultures of Politics, Politics of Culture: Re-visioning Latin American Social Movements*, ed. Sonia E. Alvarez, Evelina Dagnino, and Arturo Escobar, 1–29. Boulder, Colo.: Westview Press.

Amnesty International. 2005. "Colombia: La Ley de Justicia y Paz garantizará la impunidad para los autores de abusos contra los derechos humanos." Press release. 26 April.

Andrade, Angela, and Manuel José Amaya. 1994. "El Ordenamiento Territorial En El Instituto Geográfico 'Agustín Codazzi': Aproximación Conceptual Y Metodológica." *Revista del Proyecto SIG-PAFC* I no. 3, 32–42.

Andrews, George R. 2004. *Afro-Latin America, 1800–2000*. New York: Oxford University Press.

Anthias, Floya, and Nira Yuval-Davis. 1993. *Racialized Boundaries: Race, Nation, Gender, Colour and Class and the Anti-Racist Struggle*. London: Routledge.

Appelbaum, Nancy P. 2003. *Muddied Waters: Race, Region, and Local History in Colombia, 1846–1948*. Durham, N.C.: Duke University Press.

Appelbaum, Nancy P., Anne S. Macpherson, and Karin A. Rosemblatt, eds. 2003. *Race and Nation in Modern Latin America*. Chapel Hill: University of North Carolina Press.

Appfel Marglin, Frederique, and Stephen A. Marglin, eds. 1990. *Dominating Knowledge: Development, Culture and Resistance*. Oxford: Clarendon.

Aprile-Gniset, Jacques. 1993. *Poblamiento, hábitats y pueblos del Pacífico*. Cali: Centro Editorial, Universidad del Valle.

Arias, Andrés F. 2006. "Colombia: Ministerio de Agricultura insiste en las bondades de la ley forestal." *El Tiempo*, 28 January.

Arocha, Jaime. 1989. "Hacia una nación para los excluidos." *El Espectador*, 30 July, 14–21.

———. 1992. "Afro-Colombia Denied." *NACLA Report on the Americas* 25 no. 4, 28–31.

———. 1994. "Cultura afrocolombiana, entorno y derechos territoriales." *La política social en los 90s*, ed. Jaime Arocha, 87–105. Bogotá: Universidad Nacional de Colombia.

Arroyo, Tilson, and Yellen Aguilar Ararat. 1995. "Evaluación de sistemas productivos—una mirada al espejo." *Economías de las comunidades rurales en el Pacífico colombiano*, ed. Claudia Leal, 37–39. Bogotá: BioPacífico. Proceedings of conference held in Quibdó, 19–21 October 1994.

Asher, Kiran. 1997. "'Working from the Head Out: Revalidating Ourselves as Women, Rescuing Our Black Identity': Ethnicity and Gender in the Pacific Lowlands." *Current World Leaders* 40 no. 6, 106–27.

———. 1998. "Constructing Afro-Colombia: Ethnicity and Territory in the Pacific Lowlands." Ph.D. diss., University of Florida, Gainesville.

———. 2000. "Mobilizing the Discourses of Sustainable Economic Development and Biodiversity Conservation in the Pacific Lowlands of Colombia." *Strategies* 13 no. 1, 111–25.

———. 2001. "Meanings and Materiality: The Making of Colombian Biodiversity." Paper presented at the 23rd International Congress of the Latin American Studies Association, Washington, D.C., 6–8 September.

———. 2004. "'Texts in Context': Afro-Colombian Women's Activism in the Pacific Lowlands of Colombia." *Feminist Review* 78, 1–18.

Bacon, David. 2007. "Blood on the Palms: Afro-Colombians Fight New Plantations." *Dollars and Sense* (271), July–August, 28–34.

Balanta, Olivia, Betty Rodríguez, Sonia Sinisterra, Piedad Quiñonez, Leyla Arroyo, and Equipo Dinamizador. 1997. "Red de Mujeres Negras del Pacífico:
Tejiendo procesos organizativos autonomos." *Esteros* 9 (February), 37–42.

Barnes, Jon. 1993. "Driving Roads through Land Rights. The Colombian Plan
Pacífico." *Ecologist* 23 no. 4, 135–40.

Barriteau, Eudine. 1995. "Postmodernist Feminist Theorizing and Development
Policy and Practice in the Anglophone Caribbean: The Barbados Case."
Feminism/Postmodernism/Development, ed. Marianne H. March and Jane L.
Parpart, 142–58. New York: Routledge.

Bejarano, Ana María. 2001. "The Constitution of 1991: An Institutional Evaluation Seven Years Later." *Violence in Colombia 1990–2000*, ed. Charles Bergquist,
Ricardo Peñaranda, and Gonzalo Sánchez, 53–74. Wilmington, Del.: Scholarly Resources.

———. 2003. "Protracted Conflict, Multiple Protagonists, and Staggered Negotiations: Colombia, 1982–2002." *Canadian Journal of Latin American and
Caribbean Studies* 28 nos. 55–56, 223–48.

Bergquist, Charles W. 1986. *Coffee and Conflict in Colombia, 1886–1910*. Durham,
N.C.: Duke University Press.

———. 2001. "Waging War and Negotiating Peace: The Contemporary Cri-
sis in Historical Perspective." *Violence in Colombia 1990–2000*, ed. Charles
Bergquist, Ricardo Peñaranda, and Gonzalo Sánchez. Wilmington, Del.:
Scholarly Resources Books.

Bergquist, Charles, Ricardo Peñaranda, and Gonzalo Sánchez, eds. 2001. *Violence in Colombia 1990–2000*. Wilmington, Del.: Scholarly Resources Books.

Betancur, Ana Cecilia. 1994a. "Cómo proponen los indígenas la conformación
de sus entidades territoriales." *Esteros* 3–4 (January–March), 39–43.

———. 1994b. "Con la constitución debajo del brazo, los indígenas siguen de-
fendiendo la vida." *Esteros* 5–6 (April–September), 56–66.

Bhabha, Homi. 1994. *The Location of Culture*. New York: Routledge.

Boserup, Ester. 1970. *Women's Role in Economic Development*. New York: St.
Martin's.

Braidotti, Rosi, Ewa Charkiewicz, Sabine Häusler, and Saskia Wieringa, eds.
1994. *Women, the Environment and Sustainable Development: Towards a Theoretical Synthesis*. London: Zed Books.

Brysk, Alison. 2000. *From Tribal Village to Global Village*. Stanford, Calif.: Stanford University Press.

Bushnell, David. 1993. *The Making of Modern Colombia: A Nation in Spite of Itself*.
Berkeley: University of California Press.

Camacho, Juana. 1999. "'Todos tenemos derecho a su parte': Derechos de herencia, acceso y control de bienes en comunidades negras de la costa Pacífica
chocoana." *De montes, ríos, y ciudades: Territorios e identidades de la gente negra
en Colombia*, ed. Juana Camacho and Eduardo Restrepo, 107–30. Santa Fe de
Bogotá: Fundación Natura, Ecofondo, and ICANH.

————. 2001. "Mujeres, zoteas y hormigas arrieras: Prácticas de manejo de flora en la costa Pacífica chocoana." *Zoteas: Biodiversidad y relaciones culturales en el Chocó biogeográfico colombiano*, ed. Jesús E. Arroyo, Juana Camacho, Mireya Leyton, and Maribell González, 35–58. Bogotá: IIAP, Fundación Natura, and Fundación SWISSAID–Colombia.

————. 2004. "Silencios elocuentes, voces emergentes: Reseña bibliográfica de los estudios sobre la mujer afrocolombiana." *Panorámica afrocolombiana. Estudios sociales en el Pacífico*, ed. Mauricio Pardo, Claudia Mosquera, and María C. Ramírez, 167–212. Bogotá: ICANH and Universidad Nacional de Colombia.

Camacho, Juana., and Carlos Tapia. 1996. "Black Women and Biodiversity in the Tribuga Golf, Chocó, Colombia." Final report submitted to the MacArthur Foundation, Fundación Natura and ICANH.

Cano Correa, Claudio, and Marisol Cano Busquets. 1989. "Las etnias en la encrucijada nacional." *El Espectador*, 20 August, 3–11.

Carpay, Sander. 2004. *International Development Cooperation in the Colombia Pacific*. Utrecht: Royal Netherlands Embassy Bogotá.

Carrizosa Umaña, Julio. 1993. "El Chocó y el resto del mundo." *EcoLogica* 4 nos. 15–16, 12–17.

Casas, Fernando. 1993. "¿Porqué está 'in' el Pacíficio?" *EcoLogica* 4 nos. 15–16, 36–40.

————. 1994. "Las instituciones frente al desarrollo de la Ley 70 de 1993: Proyecto BioPacífico." *Memorias: Seminario Ley 70 de 1993 y políticas nacionales para el Pacífico colombiano*, 63–67. Bogotá: PNR.

Castro Hinestroza, Rudecindo. 1996. *Instituto de Investigaciones Ambientales del Pacifico: "John von Neumann."* Quibdó: Centro de Estudios Regionales del Pacifico.

Censat Agua Viva and Proceso de Comunidades Negras. 2007. "Convocatoria Taller Regional Monocultivo de Palma Aceitera y Agrocombustibles." Report, on file with author. http://www3.renacientes.org:8080.

Cháves Martínez, Diego A. 2005. "Agenda Pacífico XXI: Otra oportunidad para el Pacífico colombiano desaprovechada." *Economía Colombiana*, no. 311. Contraloría General de la Nación.

Cohen, Jean L. 1985. "Strategy or Identity: New Theoretical Paradigms and Contemporary Social Movements." *Social Research* 52 no. 4, 663–716.

Collins, Patricia Hill. 1991. *Black Feminist Thought: Knowledge, Consciousness, and the Politics of Empowerment*. New York: Routledge.

Colmenares, Germán. 1982. "La economia y la sociedad coloniales, 1550–1800." *Manual de Historia de Colombia*, vol. I., 225–300. Bogotá: Procultura.

Colmenares, Rafael. N.d. "El agua, un bien público. Comentarios al Proyecto de Ley del Agua y Campaña para la defensa del agua como bien público Corporación Ecofondo." Commentary, on file with author. http://www.ecofondo.org.co/ecofondo/index.php.

Comaroff, Jean, and John L. Comaroff. 2001. "Millenial Capitalism: First Thoughts on a Second Coming." *Millenial Capital and the Culture of Neo-*

liberalism, ed. Jean Comaroff and John L. Comaroff, 1–56. Durham, N.C.: Duke University Press.

Comisión Colombiana de Juristas. 2005. "Privatización de los bosques colombianos y de los territorios de las comunidades indígenas y afrocolombianas." Report, on file with author. http://www.acnur.org/.

Comité Internacional de la Cruz Roja. 2007. "Los desplazados en Colombia aumentan pese a desmovilización de 'paras'." Report, on file with author. http://www.codhes.org/.

Congreso de Colombia. 1997. "Ley para la prevención del desplazamiento forzado." *Diario Oficial*, no. 43.091, 24 July.

———. 2005. "Ley de Justicia y Paz." *Diario Oficial*, no. 45.980, 25 July.

Consolidación de la Región Amazónica. 1994–95. "Los indígenas en el marco legal." *Revista COAMA* 1.

Consultoría para los Derechos Humanos y el Desplazamiento. 2006. *Boletín de la Consultoría para los Derechos Humanos y el Desplazamiento*, no. 69. Bulletin, on file with author. http://www.codhes.org/.

Consultoría para los Derechos Humanos y el Desplazamiento and Pastoral Social. 2006. "Conclusiones problemática del desplazamiento forzado en Colombia entre 1995 y 2005." Report, on file with author. http://www.acnur.org/.

Cooper, Frederick, and Randall Packard, eds. 1997. *International Development and the Social Sciences: Essays on the History and Politics of Knowledge*. Berkeley: University of California Press.

Córdoba, Marino. 2004. "Afro-Colombians and the Colombian Civil War: A Personal Account." Lecture presented at Latin American and Caribbean Studies, International Institute, University of Michigan, Ann Arbor.

Correa, R. François.1993. "A manera de epílogo. Derechos étnicos: Derechos humanos." *Encrucijadas de Colombia Amerindia*, ed. R. François Correa, 319–34. Bogotá: ICANH and Colcultura.

———, ed. 1993. *Encrucijadas de Colombia Amerindia*. Bogotá: ICANH and Colcultura.

Cortés, Hernan. 1994. "Exposición de los representantes de Las Organizaciones de Comunidades Negras: Identidad Cultural." *Memorias: Seminario Ley 70 de 1993 y políticas nacionales para el Pacífico colombiano*, 25–27. Bogotá: PNR.

Cowen, Michael P., and Robert W. Shenton. 1996. *Doctrines of Development*. London: Routledge.

Crush, Jonathan, ed. 1995. *Power of Development*. London: Routledge.

de Friedemann, Nina S. 1985a. "Negros en Colombia: Invisibilidad y presencia." *El negro en la historia de Colombia: Fuentes orales y escritas*, 69–91. Bogotá: Fundación Colombiana de Investigaciones Folclóricas. Proceedings of conference held in Bogotá, 12–15 October 1983.

———. 1985b. "'Troncos' among Black Miners in Colombia." *Miners and Mining in the Americas*, ed. Thomas Greaves and William Culvers, 204–25. Manchester: Manchester University Press.

———. 1989. *Críele críele son. Del Pacífico negro*. Bogotá: Planeta.

de Friedemann, Nina S., and Jaime Arocha. 1984. "Estudios de negro en la antropología colombiana." *Un Siglo de investigación social: Antropología en Colombia*, ed. Jaime Arocha and Nina S. de Friedemann, 507–73. Bogotá: Etno.

———. 1986. *De sol a sol. Génesis, transformación y presencia de los negros en Colombia*. Bogotá: Planeta.

———. 1995. "Colombia." *No Longer Invisible: Afro-Latin Americans Today*, ed. Minority Rights Group, 47–76. London: Minority Rights Publications.

de Friedemann, Nina S. and Richard Cross. 1979. *Ma Ngombe: Guerros y ganaderos en Palenque*. Bogotá: Carlos Valencia Editores.

de Granda, Germán. 1988. "Los esclavos del Chocó." *Thesaurus: Boletin del Insituto Caro y Cuervo* 43 no. 1, 65–80.

de la Cadena, Marisol. 2000. *Indigenous Mestizos: The Politics of Race and Culture in Cuzco, Peru, 1919–1991*. Durham, N.C.: Duke University Press.

de Roux, Gustavo. 1992. "Historia de unas technologias inapropriadas." *Via Alterna* 1st semester, 9–13.

del Valle, Jorge I. 1995. "Ordenamiento territorial en comunidades negras del Pacífico colombiano: Olaya Herrera, Nariño." Unpublished manuscript, on file with the author.

del Valle, Jorge I., and Eduardo Restrepo, eds. 1996. *Renacientes del Guandal*. Bogotá: Proyecto Biopacífico and Universidad Nacional.

Delacour, Justin. 2000. "Plan Colombia: Rhetoric, Reality, and the Press." *Social Justice* 27 no. 4, 63.

Departamento Nacional de Planeación. 1992. "Plan Pacífico: Una nueva estrategia de desarrollo sostenible para la costa pacifica Colombiana." Approved version of document DNP-2589, March 30, Bogotá. Unpublished manuscript, on file with author.

———. 1994. "Plan para el desarrollo sustentable de la región del Pacífico colombiano." Reglamento Operativo Programa BID-Pacífico, November, Bogotá. Unpublished manuscript, on file with author.

———. 1995. "Plan de acción y programa de inversiones para el Pacifico 1995–1998." CONPES document of the Plan Pacífico, discussion version, Buenaventura, Colombia. On file with the author.

———. 2003. "Plan nacional de desarrollo 2002–2006. Hacia un estado comunitario." Bogotá: Imprenta Nacional de Colombia. http://www.presidencia .gov.co/planacio/index.htm.

———. 2007. "Estado Comunitario: Desarrollo para todos, su aplicación en el Pacífico." http://www.dnp.gov.co/.

Dover, Robert, and Joanne Rappaport. 1996. "Ethnicity Reconfigured: Indigenous Legislators and the Colombian Constitution of 1991." *Journal of Latin American Anthropology* 1 no. 2, 2–17.

Dugas, John C. 1997. "Explaining Democratic Reform in Colombia: The Origins of the 1991 Constitution." Ph.D. diss., Indiana University, Bloomington.

Echeverri, Rordrigo, and Fernando Salazar. 1994. "Zonificación ecológica del pacífico colombiano." *Revista del Proyecto SIG-PAFC* 1 no. 3, 50–60.

Eckstein, Susan, ed. 1989. *Power and Popular Protest: Latin American Social Movements*. Berkeley: University of California Press.

Edelman, Marc, and Angelique Haugerud, eds. 2005. *The Anthropology of Development and Globalization*. Malden, Mass.: Blackwell.

Escobar, Arturo. 1992. "Imagining a Post-Development Era? Critical Thought, Development and Social Movements." *Social Text* 31–32, 20–56.

———. 1995. *Encountering Development: The Making and Unmaking of the Third World*. Princeton. N.J.: Princeton University Press.

———. 1996. "Viejas y nuevas formas de capital y los dilemas de la biodiversidad." *Pacífico ¿desarrollo o diversidad? Estado, capital y movimientos sociales en el Pacífico colombiano*, ed. Arturo Escobar and Alvaro Pedrosa, 109–31. Bogotá: CEREC and Ecofondo.

———. 1997. "Cultural Politics and Biological Diversity: State, Capital, and Social Movements in the Pacific Coast of Colombia." *The Politics of Culture in the Shadow of Capital*, ed. Lisa Lowe and David Lloyd, 201–25. Durham, N.C.: Duke University Press.

———. 1998. "Whose Knowledge, Whose Nature? Biodiversity, Conservation, and the Political Ecology of Social Movements." *Journal of Political Ecology* 5, 53–82.

———. 1999. "After Nature: Steps to an Anti-Essentialist Political Ecology." *Current Anthropology* 40 no. 1, 1–30.

———. 2003. "Displacement, Development and Modernity in the Colombian Pacific." *International Social Science Journal* 175, 157–67.

Escobar, Arturo, and Sonia E. Alvarez. 1992. *The Making of Social Movements in Latin America: Identity, Strategy, and Democracy*. Boulder, Colo.: Westview Press.

Escobar, Arturo, and Alvaro Pedrosa. 1993. "¿Laboratorio para el postdesarrollo?" *Revista Universidad del Valle* 5, 34–45.

Espinosa, Monica, and Nina S. de Friedemann. 1993. "Colombia: La mujer negra en la familia y en su conceptualización." *Contribución africana a la cultura de las Américas*, ed. Astrid Ulloa, 95–111. Bogotá: ICANH.

Esteva, Gustavo. 1987. "Regenerating People's Space." *Alternatives* 12 no. 1, 125–52.

Fals Borda, Orlando. 1992. "Social Movements and Political Power in Latin America." *The Making of Social Movements in Latin America: Identity, Strategy, and Democracy*, ed. Arturo Escobar and Sonia E. Alvarez, 303–16. Boulder, Colo.: Westview Press.

Ferguson, James. 1994. *The Anti-Politics Machine: "Development," "Depoliticization," and Bureaucratic Power in Lesotho*. Minneapolis: University of Minnesota Press.

Ferguson, Kathy E. 1993. *The Man Question: Visions of Subjectivity in Feminist Theory*. Berkeley: University of California Press.

Findji, María Teresa. 1992. "From Resistance to Social Movement: The Indigenous Authorities Movement in Colombia." *The Making of New Social Movements in Latin America: Identity, Strategy, and Democracy*, ed. Arturo Escobar and Sonia E. Alvarez, 112–33. Boulder, Colo.: Westview Press.

Fiori, Lavinia. 1995. "Territorio y relaciones interétnicas en el bajo Atrato. Municipio de Río Sucio." *Esteros* 7 (August), 55–58.

Flax, Jane. 1990. *Thinking Fragments: Psychoanalysis, Feminism, and Postmodernism in the Contemporary West*. Berkeley: University of California Press.

Flórez-Flórez, Juliana. 2004. "Implosión identitaria y movimientos sociales: desafíos y logros del Proceso de Comunidades Negras ante las relaciones de género." *Conflicto e (in)visibilidad: Retos de los estudios de la gente negra en Colombia*, ed. Eduardo Restrepo and Axel Rojas, 217–44. Cauca: Editorial Universidad del Cauca.

Foucault, Michel. 1980. *Power/Knowledge: Selected Interviews and Other Writings, 1972–1977*. New York: Pantheon Books.

———. 1983. "Afterword: Subject and Power." *Michel Foucault: Beyond Structuralism and Hermeneutics*, ed. Hubert L. Dreyfus and Paul Rabinow, 208–26. Chicago: University of Chicago Press.

———. 1991. "Governmentality." *The Foucault Effect: Studies in Governmentality*, ed. Graham Burchell, Colin Gordon, and Peter Miller, 87–104. Chicago: University of Chicago Press.

Fuss, Diana. 1989. *Essentially Speaking: Feminism, Nature and Difference*. New York: Routledge.

Gallardo, Juvencio. 1986. "Colonización educativa y cultural en San Andrés Islas." *La participación del negro en la formación de las sociedades Latinoamericanas*, ed. Alexander Cifuentes, 159–66. Bogotá: Colcultura and ICANH.

García Canclini, Néstor. 1989. *Culturas híbridas: Estrategias para entrar y salir de la modernidad*. Mexico City: Grijalbo.

Gentry, Alwyn. 1993. "Sabemos más de la luna que del Chocó." *EcoLogica* 4 nos. 15–16, 56–59.

Gidwani, Vinay. 2002. "The Unbearable Modernity of 'Development'? Canal Irrigation and Development Planning in Western India." *Progress in Planning* 58, 1–80.

Gilroy, Paul. 1987. *There Ain't No Black in the Union Jack: The Cultural Politics of Race and Nation*. London: Hutchinson Press.

Giraldo, Javier. 1999. "Corrupted Justice and the Schizophrenic State in Colombia." *Social Justice* 26 no. 4, 31.

Gómez, José A. 1994. "Aserrín Aserrán, las maderas para dónde van." *Esteros* 5–6 (April–September), 23–26.

Gómez, José Luis, and Fernando Salazar. 1994. "Las instituciones frente al desarrollo de la Ley 70 de 1993: Plan de acción forestal para Colombia (PAFC)." *Memorias: Seminario Ley 70 de 1993 y políticas nacionales para el Pacífico colombiano*, 69–73. Bogotá: Plan Nacional de Rehabilitación.

Gow, David D., and Joanne Rappaport. 2002. "The Indigenous Public Voice: The Multiple Idioms of Modernity in Native Cauca." *Indigenous Movements, Self-Representation, and the State in Latin America*, ed. Kay B. Warren and Jean E. Jackson, 47–80. Austin: University of Texas Press.

Gramsci, A. 1995 (1971). *Selections from the Prison Notebooks*, ed. Quintin Hoare and Geoffrey Nowell Smith. New York: International Publishers.

Grandin, Greg. 2000. *The Blood of Guatemala: A History of Race and Nation*. Durham, N.C.: Duke University Press.

Gros, Christian. 1991. *Colombia indígena: Identidad cultural y cambio social*. Bogotá: CEREC.

Grossberg, Lawrence, Cary Nelson, and Paula A. Treichler, eds. 1992. *Cultural Studies*. London: Routledge.

Grueso, Jesus A., and Arturo Escobar. 1996. "Las cooperativas agrarias y la modernización de los agricultores." *Pacífico ¿desarrollo o diversidad? Estado, capital y movimientos sociales en el Pacífico colombiano*, ed. Arturo Escobar and Alvaro Pedrosa, 90–108. Bogotá: CEREC and Ecofondo.

Grueso, Libia R. 1993–94. "Apuntes y comentarios sobre la 3ra. Asamblea Nacional de Comunidades Negras." *Esteros* 3, 4 (October–December, January–March), 32–43.

———. 1998. "El ejercicio del derecho al territorio en la comunidad negra del Pacífico Sur: Un reto en doble vía." Buenaventura, Colombia. Unpublished manuscript, on file with author.

Grueso, Libia R., and Leyla A. Arroyo. 2002. "Women and the Defense of Place in Colombian Black Movement Struggles." *Development* 45 no. 1, 67–73.

Grueso, Libia R., Carlos Rosero, and Arturo Escobar. 1998. "The Process of Black Community Organizing in the Southern Pacific Coast Region of Colombia." *Cultures of Politics/Politics of Cultures: Re-visioning Latin American Social Movements*, ed. Sonia E. Alvarez, Evelina Dagnino, and Arturo Escobar, 196–219. Boulder, Colo.: Westview Press.

Guhl Nannetti, Ernesto. 2006. "Con más arrogancia que racionalidad. Colombia: Ministros de Agricultura y Medio Ambiente tratan la ley forestal." *El Tiempo*, 18 January.

Gupta, Akhil. 1998. *Postcolonial Developments: Agriculture in the Making of Modern India*. Durham, N.C.: Duke University Press.

Gutiérrez de Pineda, Virginia. 1994 (1968). *Familia y cultura en Colombia: Tipologías, funciones y dinámica de la familia. Manifestaciones múltiples a través del mosaico cultural y estructuras sociales*. Medellín: Editorial Universidad de Antioquia.

Haas, Peter M., Marc A. Levy, and Edward A. Parson. 1992. "Appraising the Earth Summit: How Should We Judge UNCED's Success?" *Environment* 34 no. 8, 5–15, 26–36.

Hall, Stuart. 1986. "Gramsci's Relevance for the Study of Race and Ethnicity." *Journal of Communication Inquiry* 10 no. 3, 5–27.

———. 1988. "The Toad in the Garden: Thatcherism among the Theorists." *Marxism and the interpretation of culture*, ed. C. Nelson, G. Lawrence, 35–73. Urbana: University of Illinois Press.

———. 1992a. "New Ethnicities." *"Race," Culture, and Difference*, ed. James Donald and Ali Rattansi, 252–59. London: Sage Publishers and Open University.

———. 1992b. "What Is This 'Black' in Black Popular Culture?" *Black Popular Culture*, ed. Gina Dent, 21–33. Seattle: Bay Press.

Haraway, Donna J., ed. 1991. *Simians, Cyborgs, and Women: The Reinvention of Nature*. New York: Routledge.

Harvey, David. 1989. *The Condition of Postmodernity*. Malden, Mass.: Blackwell.

Hoffman, Odile. 1999. "Territorialidades y alianzas: Construcción y activación de espacios locales en el Pacífico." *De montes, ríos, y ciudades: Territorios e identidades de la gente negra en Colombia*, ed. Juana Camacho and Eduardo Restrepo. Santa Fe de Bogotá: Fundación Natura, Ecofondo, and ICANH.

———. 2002. "Collective Memory and Ethnic Identities in the Colombian Pacific." *Journal of Latin American Anthropology* 7 no. 2, 118–39.

hooks, bell. 1990. *Yearning: Race, Gender, and Cultural Politics*. Boston: South End Press.

———. 1992. *Black Looks: Race and Representation*. Boston: South End Press.

Hudson, Randall O. 1964. "The Status of the Negro in Northern South America, 1820–1860." *Journal of Negro History* 49 no. 4, 225–39.

Human Rights Watch. 2005. "Internal Displacement. Colombia: Displaced and Discarded. The Plight of Internally Displaced Persons in Bogotá and Cartagena." *Human Rights Watch* 17 no. 4(B).

Hurtado, María Lucia. 1996. "La construcción de una nación multiétnica y pluricultural." *Pacífico ¿desarrollo o diversidad? Estado, capital y movimientos sociales en el Pacífico colombiano*, ed. Arturo Escobar and Alvaro Pedrosa, 329–52. Bogotá: CEREC and Ecofondo.

Instituto Geográfico Agustín Codazzi (IGAC). 2001. *Ordenamiento Territorial*. Report, on file with author. http://www.igac.gov.co.

Inter-American Foundation. 2001. "Economic Development in Latin American Communities of African Descent." *Panel Presentations at the XXIII International Congress of the Latin American Studies Association*, 6–8 September. Washington, D.C.: Inter-American Foundation.

Jackson, Jean E. 2002. "Caught in the Crossfire: Colombia's Indigenous Peoples during the 1990s." *The Politics of Ethnicity: Indigenous Peoples in Latin American States*, ed. David Maybury-Lewis, 107–33. Cambridge, Mass.: David Rockefeller Center for Latin American Studies, Harvard University Press.

Jaquette, Jane S., ed. 1994. *The Women's Movement in Latin America: Participation and Democracy*. Boulder, Colo.: Westview Press.

Jaramillo Uribe, Jaime. 1963. "Esclavos y señores en la sociedad colombiana del siglo XVIII." *Anuario Colombiano de Historia Social y de la Cultura* 1, no. 1, 5–76

———. 1989. "Los estudios afroamericanos y afrocolombianos: Balance y perspectivas." *Ensayos de historia social*, vol. 2, 203–224. Bogotá: Tercer Mundo Editores.

Jelin, Elizabeth. 1990. *Women and Social Change in Latin America*. London: Zed Books.

Jiménez, Michael F. 1995. "At the Banquet of Civilization: The Limits of Planter Hegemony in Early-Twentieth-Century Colombia." *Coffee, Society, and Power in Latin America*, ed. William Roseberry, Lowell Gudmundson, and Mario Samper Kutschbach. Baltimore: Johns Hopkins University Press.

Kabeer, Naila. 1994. *Reversed Realities: Gender Hierarchies in Development Thought.* London: Verso.

Keck, Margaret E., and Kathryn Sikkink. 1998. *Activists beyond Borders: Advocacy Networks in International Politics.* Ithaca, N.Y.: Cornell University Press.

Khittel, Stefan R. F. 2001. "Usos de la historia y historiografía por parte de las ONG y Organizaciones de Bases de las comunidades negras en el Chocó." *Acción colectiva, estado y etnicidad en el Pacífico colombiano*, ed. Mauricio Pardo, 71–92. Bogotá: ICANH and Colciencias.

Klinger, William. 1996. "Plan de desarrollo de comunidades negras." *Comunidades negras, territorio y desarrollo: Propuestas y discusion*, 157–70. Medellín: Esteros, Edición Especial.

Laclau, Ernesto, and Chantal Mouffe. 1985. *Hegemony and Socialist Strategy: Towards a Radical Democratic Politics.* London: Verso.

Leal, Claudia. 2003. "Mapping the Colombian Conflict." *Canadian Journal of Latin American and Caribbean Studies* 28 nos. 55–56, 211–22.

LeGrand, Catherine C. 2003. "The Colombian Crisis in Historical Perspective." *Canadian Journal of Latin American and Caribbean Studies* 28 nos. 55–56, 165–209.

Lele, Sharadchandra M. 1991. "Sustainable Development: A Critical Review." *World Development* 19 no. 6, 607–21.

Leys, Colin. 1996. *The Rise and Fall of Development Theory.* Oxford: James Currey.

Leys Stepan, Nancy. 1991. *"The Hour of Eugenics": Race, Gender, and Nation in Latin America.* Ithaca, N.Y.: Cornell University Press.

Lind, Amy. 2005. *Gendered Paradoxes: Women's Movements, State Restructuring, and Global Development in Ecuador.* University Park: Pennsylvania State University Press.

Llano, María Clara. N.d. *La gente del Bajo Patía: Un estudio de historia regional en el Pacífico colombiano.* Bogotá: ICANH, PNR, and Junta Pro-Defensa de los Patías.

López, Andrés. 1999. "The Implementation of the Cartagena Agreement: Some Thoughts and Concerns." *Más Allá del Derecho* 6(18–19), 83–93.

López, Manuel E. 1996. "Estado actual de la explotaciones madereras en le Bajo Atrato Chocoano." *Esteros* 4 no. 8, 28–37.

Lorde, Audre. 1992. "Age, Race, Class and Sex: Women Redefining Difference." *Knowing Women: Feminism and Knowledge*, ed. Helen Crowley and Susan Himmelweit, 47–54. Cambridge: Polity Press and the Open University.

Losonczy, Anne-Marie. 1991. "El luto de sí mismo. Cuerpo, sombra y muerte entre los negros-colombianos del Chocó." *America Negra* 1, 43–61.

———. 1993. "De lo vegetal a lo humano: Un modelo cognitivo afro-colombiano del Pacífico." *Revista Colombiana de Antropología* 30, 39–57.

———. 1999. "Memorias e identidad: Los negro-colombianos del Chocó." *De montes, ríos, y ciudades: Territorios e identidades de la gente negra en Colombia*, ed. Juliana Camacho and Eduardo Restrepo, 13–24. Santa Fe de Bogotá: Fundación Natura, Ecofondo, and ICANH.

Lowe, Lisa, and David Lloyd. 1997. "Introduction." *The Politics of Culture in the Shadow of Capital*, ed. Lisa Lowe and David Lloyd, 1–32. Durham, N.C.: Duke University Press.

Lozano, Betty Ruth. 1992. "Una crítica a la sociedad occidental patriacal y racista desde la perspectiva de la mujer negra." *Pasos* 42 (July–August), 11–20.

———. 1996. "Mujer y desarrollo." *Pacífico ¿desarrollo o diversidad? Estado, capital y movimientos sociales en el Pacífico colombiano*, ed. Arturo Escobar and Alvaro Pedrosa, 176–204. Bogotá: CEREC and Ecofondo.

McClintock, Anne, Aamir Mufti, and Ella Shohat, eds. 1997. *Dangerous Liaisons: Gender, Nation, and Postcolonial Perspectives*. Minneapolis: University of Minnesota Press.

McMichael, Philip. 2000. "Globalization: Myths and Realities." *From Modernization to Globalization: Perspectives on Development and Change*, ed. J. Timmons Roberts and Amy Bellone Hite, 274–94. Malden, Mass.: Blackwell.

Mina, María del Rosario. 1995. "La cuestion de definir lineamientos para politicas del desarrollo de la mujer del Pacífico. Desde las realidades de la mujer de Buenaventura." Unpublished manuscript, on file with author.

Mingorance, Fidel, Flamina Minelli, and Hélène Le Du. 2004. "El cultivo de la palma africana en el Chocó. Legalidad ambiental, territorial y derechos humanos." Report, on file with author. http://www.acnur.org/.

Ministerio de Ambiente Vivienda y Desarrollo Territorial. 2001. "Ecorregiones estratégicas regionales. Región de concertación SINA—Región Pacífico." Report, on file with author. http://www.minambiente.gov.co/.

———. 2006. "Bienestar de la población afrocolombiana en la mira del gobierno." *Noticias*, 16 December. Press release. http://www.minambiente.gov.co/.

Mitchell, Timothy. 1991. "America's Egypt: Discourse of the Development Industry." *Middle Eastern Report* 29 (March–April), 18–36.

Mohammed, Patricia, and Catherine Shepherd, eds. 1988. *Gender in Caribbean Development*. Jamaica: Canoe Press, University of the West Indies.

Mohanty, Chandra Talpade. 1997. "Women Workers and Capitalist Scripts: Ideologies of Deomination, Common Interests and the Politics of Solidarity." *Feminist Genealogies, Colonial Legacies, Democratic Futures*, ed. M. Jacqui Alexander, 3–29. New York: Routledge.

———. 2003. "'Under Western Eyes' Revisited: Feminist Solidarity through Anticapitalist Struggles." *Feminism without Borders: Decolonizing Theory, Practicing Solidarity*, ed. Chandra Talpade Mohanty, 221–51. Durham, N.C.: Duke University Press.

Mohanty, Chandra Talpade, Ann Russo, and Lourdes Torres, eds. 1991. *Third World Women and the Politics of Feminism*. Bloomington: Indiana University Press.

Monje, César, and Ricardo Castillo. 1995. "Zonificación Ecologica del Pacífico," Paper presented at *Taller sobre constucción del territorio en el Pacífico y Ley 70*, August 24–26, at the Universidad del Valle, Cali.

Moraga, Cherríe, and Gloria Anzaldúa. 1983. *This Bridge Called My Back: Writings by Radical Women of Color*. New York: Kitchen Table/Women of Color Press.

Moser, Caroline O. N. 1993. *Gender Planning and Development*. London: Routledge.

Motta, Nancy. 1992. "Grupos negros del Pacífico." *Diversidad es riqueza: Ensayos sobre la realidad colombiana*, 95–100. Bogotá: ICANH and Consejería Presidencial para los Derechos Humanos.

———. 1994. "Identidad étnica, género y familía en la cultura negra del pacífico colombiano." Paper presented at the Congreso Latinoamericano de Familia, Siglo XXI, Medellín, 1994.

Mufti, Aamir, and Ella Shohat. 1997. "Introduction." *Dangerous Liasons: Gender, Nation, and Postcolonial Perspectives*, eds. Anne McClintock, Aamir Mufti, and Ella Shohat, 1–11. Minneapolis: University of Minnesota Press.

"Mujeres, . . . con olor y sabor a Chiyangua: Una entrevista." 1998. *El Hilero*, 15–17.

"Mujeres Negras e Indigenas Definiendo Futuro." 1997. *Esteros* 5 no. 9, Special Issue.

"Mujeres negras, Latinoamerica." 1995. *Fempress*, special issue.

Múnera, Luis Fernando. 1994. "Las instituciones frente al desarrollo de la Ley 70 de 1993: INCORA." *Memorias: Seminario Ley 70 de 1993 y politicas nacionales para el Pacífico colombiano*. Bogotá: PNR.

Murillo, Pastor. 1994. "Los paros cívicos en el Pacífico colombiano." *Esteros* 5–6 (April–September), 34–37.

Nanda, Meera. 2002. "Do the Marginalized Valorize the Margins? Exploring the Dangers of Difference." *Feminist Post-Development Thought: Rethinking Modernity, Post-Colonialism and Representation*, ed. Kriemild Saunders, 212–23. London: Zed Books.

Ng'weno, Bettina. 2000. *On Titling Collective Property, Participation, and Natural Resource Management: Implementing Indigenous and Afro-Colombian Demands, a Review of Bank Experience in Colombia*. Washington, D.C.: World Bank.

———. 2003. "Indigeneity, Territory and State: The Effects of Imperial Remnants in Afro-Colombian Territorial Claims." Paper presented at the American Anthropological Society Meetings, 19–23 November, Chicago.

———. 2007. *Turf Wars: Territory and Citizenship in the Contemporary State*. Stanford, Calif.: Stanford University Press.

Offen, Karl H. 2003. "The Territorial Turn: Making Black Territories in Pacific Colombia." *Journal of Latin American Geography* 2 no. 1, 43–73.

Olander, Jacob. 1995. "¿Perdidas a la lata? La economía de la explotación de Palma de Naidi en el Bajo Anchicayá." *Economías de las comunidades rurales en el Pacífico colombiano*, ed. Claudia Leal, 1113–20. Bogotá: BioPacífico. Proceedings of conference held in Quibdó, 19–21 October 1994.

Ong, Aihwa. 1987. *Spirits of Resistance and Capitalist Discipline: Factory Women in Malaysia*. Albany: State University of New York Press.

————. 1988. "Colonialism and Modernity: Feminist Re-presentations of Women in Non-Western Societies." *Inscriptions* 3–4 no. 2, 79–93.

Organización de Comunidades Negras. 1996. "Movimiento negro, identidad y territorio: Entrevista con la Organización de Comunidades Negras de Buenaventura." *Pacífico ¿desarrollo o diversidad? Estado, capital y movimientos sociales en el Pacífico colombiano*, ed. Arturo Escobar and Alvaro Pedrosa, 245–65. Bogotá: CEREC and Ecofondo.

Oslender, Ulrich. 1999. "Espacio e identidad en el Pacífico colombiano." *De montes, ríos, y ciudades: Territorios e identidades de la gente negra en Colombia*, ed. J. Camacho and Eduardo Restrepo, 25–48. Santa Fe de Bogotá: Fundación Natura, Ecofondo, and ICANH.

————. 2001. "La lógica del río: Estructuras espaciales del proceso organizativo de los movimientos sociales de comunidades negras en Pacífico colombiano." *Acción colectiva, estado y etnicidaden el Pacífico colombiano*, ed. Mauricio Pardo. Bogotá: ICANH and Colciencias.

Pacheco, Esperanza, and Alberto Achito. 1993. "Relaciones interétnicas en el Chocó de hoy." *Contribución africana a la cultura de las Américas*, ed. A. Ulloa, 217–23. Bogotá: ICANH.

Pacheco, Esperanza, and Jairo Jaime Velásquez Zamudio. 1993. "Relaciones interétnicas de los Emberá del Bajo Chocó." *Encrucijadas de Colombia Amerindia*, ed. François Correa Rubio., 269–85. Bogotá: ICANH and Colcultura.

Palacios, Marco. 1980. *Coffee in Colombia, 1850–1970: An Economic, Social and Political History*. Cambridge: Cambridge University Press.

————. 2006. *Between Legitimacy and Violence: A History of Colombia, 1875–2002*. Durham, N.C.: Duke University Press.

Pardo, Mauricio. 1996. "Movimientos sociales y relaciones interetnicas." *Pacífico ¿desarrollo o diversidad? Estado, capital y movimientos sociales en el Pacífico colombiano*, ed. Arturo Escobar and Alvaro Pedrosa, 299–315. Bogotá: CEREC and Ecofondo.

————. 1998. "Construcción reciente de elementos de liderazgo en el Pacífico colombiano." *Modernidad, identidad y desarrollo: Construcción de sociedad y recreación cultural en contextos de modernización*, ed. María L. Sotomayor, 51–72. Bogotá: ICANH.

Pardo, Mauricio, and Manuela Álvarez. 2001. "Estado y movimiento negro en el Pacífico colombiano." *Acción colectiva, estado y etnicidad en el Pacífico colombiano*, ed. Mauricio Pardo, 229–58. Bogotá: ICANH and Colciencias.

Pedraza Gómez, Sandra. 1986. "Para una investigación sobre la nacionalización del archipielago de San Andrés y Providencia." *La participación del negro en la formación de las sociedades Latinoamericanas*, ed. Alexander Cifuentes, 131–42. Bogotá: Colcultura and ICANH.

Pedrosa, Álvaro. 1996. "La institucionalizacion del desarrollo." *Pacífico ¿desarrollo o diversidad? Estado, capital y movimientos sociales en el Pacífico colombiano*, ed. Arturo Escobar and Alvaro Pedrosa, 66–89. Bogotá: CEREC and Ecofondo.

Pedrosa, Álvaro, Alfredo Vanín, and Nancy Motta. 1994. *La Vertiente Afropací-fica de la Tradición Oral: Géneros y catalogación.* Cali: Universidad del Valle and Proyecto Opción Pacífico.

Peet, Richard, and Michael Watts. 1996. "Liberation Ecology: Development, Sustainability, and Environment in an Age of Market Triumphalism." *Liberation Ecologies: Environment, Development, Social Movements,* ed. Richard Peet and Michael Watts, 1–45. London: Routledge.

Perea, Berta I. 1986. "La familia afrocolombiana del Pacífico." *La participación del negro en la formación de las sociedades latinoamericanas,* ed. Alexander Cifuentes, 117–30. Bogotá: Colcultura and ICANH.

———. 1990. "Estructura familiar afrocolombiana: Elementos que definen la estructura familiar de descendientes de africanos nacidos en Colombia." Hegoa Working Paper no. 5, Bilbao.

Peterson, Sarah. 2002. "People and Ecosystems in Colombia: Casualties of the Drug War." *Independent Review* 6 no. 3, 427–40.

Piedrahíta, Diego, and María Estella Pineda. 1993. "En vías de desarrollo: Lista de proyectos de desarrollo para el Pacífico." *EcoLogica* 4 nos. 15–16, 83–86.

Proceso de Comunidades Negras. 2006. "Declaración de los Procesos de Comunidades Negras en Colombia y demás organizacions suscribientes, en el marco del Encuentro y Movilización por la Vida, la Alegría, la Esperanza y la Libertad de los Pueblos." Acompaz. Declaration, on file with author. http://www.acompaz.org.

———. N.d. "El Proceso de comunidades negras en el centro-sur del Pacífico y el problema del 'desarrollo.'" Buenaventura, Colombia.Unpublished manuscript, on file with author.

Proceso de Comunidades Negras and Organización Regional Emberá-Waunana del Chocó. 1995. *Territorio, etnia, cultura e investigación en el Pacífico colombiano.* Cali: Fundación Habla/Scribe.

Proyecto BioPacífico. 1994. "Conservación de la biodiversidad en el Chocó biogeografico: Plan Operativo Anual, Marzo 1994–Febrero 1995." Ministerio de Medio Ambiente, Global Environmental Facility and the United Nations Development Programme. Unpublished manuscript, on file with author.

Proyecto Ríos Vivos. 2000. "Proyecto Ríos Vivos: Informe del Proyecto Final." Cali: Ecofondo, Fundación Habla/Scribe, and Red Matamba y Guasá.

Rahnema, Majid, and Victoria Bawtree, eds. 1997. *The Post-Development Reader.* London: Zed Press.

Ramírez, María Clemencia. 2002. "The Politics of Identity and Cultural Difference in the Colombian Amazon: Claiming Indigenous Rights in the Putumayo Region." *The Politics of Ethnicity: Indigenous Peoples in Latin American States,* ed. David Maybury-Lewis, 135–66. Cambridge, Mass.: David Rockefeller Center for Latin American Studies, Harvard University Press.

Rappaport, Joanne, and Robert Dover. 1996. "Ethnicity Reconfigured: Indigenous Legislators and the Colombian Constitution of 1991: Introduction." *Journal of Latin American Anthropology* 1 no. 2, 2–17.

Red Matamba y Guasá. 1997. *Segundo Encuentro: Taller Sub-Regional de Orga-nizaciones de Mujeres del Pacífico Caucano 'Matamba y Guasá': Fuerza y Con-vocatoria de la Mujer del Pacífico Caucano.* Timbiquí y Guapi, Cauca: Fun-dación Chiyangua de Guapi, Grupo de Promoción de Santa Rosa de Saija, Asociación Apoyo a la Mujer de Timbiquí, and Asociación Manos Negras de Guapi.

———. 1997–98. "Visión y perspectiva política y organizativa de la Red de Or-ganizaciones de Mujeres de Cauca: La Red es una familia numerosa." *Boletin Anual de Matamba y Guasá*, 13–17.

Redclift, Michael. 1987. *Sustainable Development: Exploring the Contradictions.* London: Metheun.

Republic of Colombia, Ministerio de Medio Ambiente, and Proyecto BioPací-fico. 1994. "Politica de conservación de la biodiversidad para el desarrollo sostenible de la region biogeografica del Pacífico." Draft proposal, Bogotá.

Restrepo, Eduardo. 1996. "Cultura y biodiversidad." *Pacífico ¿desarrollo o diver-sidad? Estado, capital y movimientos sociales en el Pacífico colombiano*, ed. Arturo Escobar and Alvaro Pedrosa, 220–41. Bogotá: CEREC and Ecofondo.

———. 1998. "La construcción de la etnicidad: Comunidades negras en Colom-bia. *Modernidad, identidad y desarrollo: Construcción de sociedad y re-creación cultural en contextos de modernización*, ed. María L. Sotomayor, 341–58. Bo-gotá: ICANH.

———. 1999. "Territorios e identidades híbridas." *De montes, ríos, y ciudades: Territorios e identidades de la gente negra en Colombia*, ed. Joanne Camacho and Eduardo Restrepo, 221–44. Santa Fe de Bogotá: Fundación Natura, Ecofondo, and ICANH.

———. 2001. "Imaginando comunidad negra: Etnografía de la etnización de poblaciones negras en el Pacífico sur colombiano." *Acción colectiva, estado y etnicidad en el Pacífico colombiano*, ed. Mauricio Pardo, 41–70. Bogotá: ICANH and Colciencias.

———. 2004. "Ethnicization of Blackness in Colombia." *Cultural Studies* 18 no. 5, 698–715.

Restrepo, Mónica H. 1992. "Poblamiento y estructura social de las comunidades negras del medio Atrato." Undergraduate thesis, Department of Sociology, Universidad Nacional de Colombia, Bogotá.

Richani, Nazih. 2002. *Systems of Violence: The Political Economy of War and Peace in Colombia.* Albany: State University of New York Press.

———. 2002. "Colombia at the Crossroads. The Future of the Peace Accords: Report on Post Cold War Latin America." *NACLA Report on the Americas* 35 no. 4, 17–21.

Ríos, Germán. 1993a. "Madera va, madera viene: Debate forestal en el Atrato." *Esteros* 2 (April–June), 45–48.

———. 1993b. "Riquezas y retórica en el Pacífico, el caso del bajo Atrato." *Este-ros* 1 (January–March), 20–27.

———. 1994. "Codechocó y tríplez pizano en el rio Atrato: Un dueto colonial." *Esteros* 3–4 (January–March), 44–45.

Roa Avendaño, Tatiana. 2007. "Colombia's Palm Oil Biodiesel Push." Washington D.C.: International Relations Center, Americas Policy Program. http:// americas.irc-online.org/.

Rodríguez, Betty. 1997. "Visión y perspectiva político organizativa de la Red de Mujeres Negras." *Esteros* 9 (February), 33–35.

Rodrik, Dani. 2000. "Has Globalization Gone Too Far?" *From Modernization to Globalization: Perspectives on Development and Change*, ed. J. Timmons Roberts and Amy Bellone Hite, 298–305. Malden, Mass.: Blackwell.

Rojas, Jeannette. S. 1996. "Las mujeres en moviemiento: Cronica de otras miradas." *Pacífico ¿desarrollo o diversidad? Estado, capital y movimientos sociales en el Pacífico colombiano*, ed. Arturo Escobar and Alvaro Pedrosa, 205–18. Bogotá: CEREC and Ecofondo.

Roldán, Roque. 1993. Reconocimiento legal de tierras a indígenas en Colombia." *Reconocimiento y demarcación de territorios indigenas en la Amazonia*, 56–88. Bogotá: Fundación Gaia and CEREC.

Romero, Mario D. 1990–91. "Procesos de poblamiento y organización social en la costa Pacífica colombiana." *Anuario Colombiano de Historia Social y de la Cultura* 18–19, 9–31.

———. 1995. "La familia negra en los siglos XVIII y XIX en la costa Pacífica: Hipótesis sobre continuidades y cambios en el siglo XX." Paper presented at Estado del Arte Sobre Estudios de Familia Negra en la Sociedad Colombiana, 4–5 August, Cali.

Roseberry, William. 1989. *Anthropologies and Histories: Essays in Culture, History and Political Economy*. New Brunswick, N.J.: Rutgers University Press.

Rosero, Carlos. 1993. "Las comunidades afroamericanas y la lucha por el territorio." *Esteros* 1 (January–March), 28–31.

———. 1995. "Ruidos internos de la dinamica de las organizaciones de comunidades negras." Paper presented at the Taller sobre Construcción del Territorio en el Pacífico y Ley 70, Cali. 24–26 August. Cali: Universidad del Valle.

———. 1996. "Reflexiones sobre el concepto de desarrollo entre comunidades negras." *Comunidades Negras, territorio y desarrollo: Propuestas y discusion*, 179–86. Medellín: Esteros, Edición Especial.

———. 2002. "Los Afrodescendientes y el conflicto armado en Colombia: La insistencia en lo propio como alternativa." *Afrodescendientes en las Américas: Trajectorias sociales e identitarias, 150 años de la abolición de la esclavitud en Colombia*, ed. Claudia Mosquera, Mauricio Pardo, and Odile Hoffman, 547–59. Bogotá: Universidad Nacional de Colombia and ICANH.

———. 2007. "¡Nadie se salva solo y nadie salva a nadie!" Centro de Medios Independientes de Colombia. On file with author. http://colombia.indymedia .org/.

Rostow, Walt W. 1960. *The Stages of Economic Growth: A Non-Communist Manifesto*. Cambridge: Cambridge University Press.

Rowe, William, and Vivian Schelling. 1993. *Memory and Modernity: Popular Culture in Latin America*. London: Verso.

Ruiz, Juan Pablo. 1993. "GEF: Donación o evasión." *EcoLogica* 4 nos. 15–16, 32–35.

Ruiz Serna, Daniel. 2007. "La palma de aceite y la usurpación de territorio a las comunidades negras." Centro de Investigación y Educación Popular (CINEP), 7 March. Report, on file with author. http://www.piedadcordoba.net/.

Rutherford, Jonathan. 1990. "A Place Called Home: Identity and the Cultural Politics of Difference." *Identity: Community, Culture, Difference*, ed. Jonathan Rutherford, 9–27. London: Lawrence and Wishart.

Sachs, Wolfgang, ed. 1992. *The Development Dictionary: A Guide to Knowledge as Power*. London: Zed Books.

Safa, Helen I. 1990. "Women's Social Movements in Latin America." *Gender and Society* 4 no.3, 354–69.

Safford, Frank, and Marco Palacios. 2002. *Colombia: Fragmented Land, Divided Society*. New York: Oxford University Press.

Sala, Mariella. 1995. "Mujeres negras de Ecuador: Entrevista a Norma Rodríguez." *Fempress* (special issue), 38–39.

Salazar, Fernando. 1994. "Las instituciones frente al desarrollo de la Ley 70 de 1993: Zonificación ecológica del Pacífico Colombiano, plan de acción forestal para Columbia (PAFC)." *Memorias: Seminario Ley 70 de 1993 y políticas nacionales para el Pacífico Colombiano*. Bogotá: Plan Nacional de Rehabilitación.

Salinas, Yamile. 2005. "La Palma de Aceite: Desarrollo y derechos étnicos 2005." *El Tiempo*, 11 January.

Sánchez, Enrique. 1995. "Del extractivismo a las etnoagriculturas: Las miradas sobre la economía de las comunidades rurales negras e indígenas en el Pacífico." *Memorias del foro: Las economías rurales indígenas, negras y mestizas en el Pacífico colombiano*, 15–34. Bogotá: Proyecto BioPacífico.

Sánchez, Enrique, and Roque Roldán. 2002. *Titulación de los territorios comunales afrocolombianos e indígenas en la costa pacífica de Colombia*. Washington, D.C.: World Bank.

Sánchez, Enrique, Roque Roldán, and María Fernanda Sánchez. 1993. *Derechos e identidad: Los pueblos indígenas y negros en la Constitución Política de Colombia de 1991*. Bogotá: Consolidación de la Amazonía and European Union. Santa fe de Bogotá: Disloque Editores.

Sanders, James E. 2004. *Contentious Republicans: Popular Politics, Race, and Class in Nineteenth-Century Colombia*. Durham, N.C.: Duke University Press.

Sanders, Thomas G. 1982. *Colombia's Chocó: The Inertia of Underdevelopment*. Hanover, N.H.: Universities Field Staff International.

Sarmiento Gómez, Alfredo, Lucía Mina Rosero, Carlos A. Malaver, and Sandra Álvarez Toro. 2003. "Colombia: Human Development Progress towards the Millennium Development Goals." Background papers, Human Development Report 2003, United Nations Development Programme.

Schild, Veronica. 1998. "New Subjects of Rights? Women's Movements and the Construction of Citizenship in the 'New Democracies.'" *Cultures of Politics/Politics of Cultures: Re-visioning Latin American Social Movements*, ed. Sonia

E. Alvarez, Evelina Dagnino, and Arturo Escobar, 93–117. Boulder, Colo.: Westview Press.

Scott, Joan W. 1988. "Gender: A Useful Category of Historical Analysis." *Gender and the Politics of History*, 28–50. New York: Columbia University Press.

———. 1990. "Deconstructing Equality-Versus-Difference." *Conflicts in Feminism*, ed. M. Hirsch and E. Fox-Keller, 134–48. London: Routledge.

———. 1992. "Experience." *Feminists Theorize the Political*, ed. Judith Butler and Joan W. Scott, 22–40. London: Routledge.

Sen, Gita, and Caren Grown. 1987. *Development, Crises, and Alternative Visions: Third World Women's Perspectives*. New York: Monthly Review Press.

Sheahan, John. 1987. *Patterns of Development in Latin America: Poverty, Repression, and Economic Strategy*. Princeton, N.J.: Princeton University Press.

Sheth, D. L. 1987. "Alternative Development as Political Practice." *Alternatives* 12 no. 2, 155–71.

Shiva, Vandana. 1988. *Staying Alive: Women, Ecology, and Survival*. London: Zed Books.

———. 1990. "Biodiversity, Biotechnology and Profit: The Need for a People's Plan to Protect Biological Diversity." *Ecologist* 20 no. 2, 44–47.

Slater, David, ed. 1985. *New Social Movements and the State in Latin America*. Latin American Studies no. 29. Amsterdam: Center for Latin American Research and Documentation.

Spivak, Gayatri C. 1989. "The Political Economy of Women as Seen by a Literary Critic." *Coming to Terms: Feminism, Theory, Politics*, ed. Elizabeth Weed, 218–29. New York: Routledge.

———. 1990. *The Post-Colonial Critic: Interviews, Strategies, Dialogues*, ed. Sarah Harasym. New York: Routledge.

———. 1999. *A Critique of Postcolonial Reason: Toward a History of the Vanishing Present*. Cambridge, Mass.: Harvard University Press.

Taussig, Michael T. 1980. *The Devil and Commodity Fetishism in South America*. Chapel Hill: University of North Carolina Press.

Thayer, Millie. 2001. "Transnational Feminism: Reading Joan Scott in the Brazilian Sertão." *Ethnography* 2 no. 2, 243–27.

Thorne, Eva T. 2001. "The Politics of Afro-Latin American Land Rights." Paper presented at the 23d International Congress of the Latin American Studies Association, Washington, D.C., 6–8 September.

Tinker, Irene, ed. 1990. *Persistent Inequalities: Women and World Development*. Oxford: Oxford University Press.

Triana, Adolfo. 1993. "Grupos étnicos y Constitución." *Contribución africana a la cultura de las Américas*, ed. Astrid Ulloa, 195–200. Bogotá: ICANH.

Tulio Diaz, Servio. 1993. "Proyecto Biopacífico." *EcoLogica* 4 nos., 15–16, 24–31.

United Nations High Commissioner for Refugees. 2006. "La reglamentación de la 'Ley de Justicia y Paz' no logra establecer adecuadamente el respeto por los derechos de las víctimas," 4 January, Bogotá. Press release, on file with author. http://www.hchr.org.co/.

Urrea Giraldo, Fernando, Hector F. Ramírez, and Carlos Viáfara López. 2001. *Perfiles sociodemográficos de la población afrocolombiana en contextos urbano-regionales del país a comienzos del siglo XXI*. Cali: Centro de Investigaciones y Documentación Socioeconómica, Institut de Recherche pour le Développement, and Colciencias.

Van Cott, D. L. 2000. *The Friendly Liquidation of the Past: The Politics of Diversity in Latin America*. Pittsburg: University of Pittsburg Press.

Vargas Sarmiento, Patricia. 1993. *Los embera y los cuna: Impacto y reacción ante la ocupación española siglos XVI y XVII*. Bogotá: CEREC and ICANH.

Vásquez, Édgar. 1994. "Desarrollo del Pacífico colombiano en el marco de la internacionalización de la economía y la Constitución de 1991." *Comunidades negras: Territorio, identidad y desarrollo*, 51–69. Bogotá: PNR, ICANH, Colcultura, and UNDP.

Vásquez, Miguel A. 1994. *Las caras lindas de ni gente negra: Legislacion historica para las comunidades negras de Colombia*. Bogotá: ICANH, Colcultura, PNR, and UNDP.

Velasco, Alvaro C., and Juan C. Preciado. 1994. "Elecciones y minorías: Retos a la diversidad cultural." *Esteros* 5–6 (April–September), 27–33.

Vélez, Germán. 2005. "El Congreso de la República, aprueba la ley forestal." Grupo Semillas. On file with author. www.semillas.org.co/.

Villa, William. 1994. "San Andrés y Providencia. Entre la destrucción ecológica y cultural." *Esteros* 3–4 (January–March), 27–31.

———. 1995. "El plan de desarrollo de comunidades negras en la ley 70 de 1993." *Esteros* 7 (August), 36–40.

———. 1998. "Movimiento social de comunidades negras en el Pacífico colombiano. La construcción de una noción de territorio y región." *Geografía humana de Colombia*, vol. 6, ed. Adriana Maya Restrepo, 431–49. Bogotá: Instituto Colombiano de Cultura Hispánica.

———. 2001. "La sociedad negra del Chocó." *Acción colectiva, estado y etnicidad en el Pacífico colombiano*, ed. Mauricio Pardo, 207–28. Bogotá: ICANH and Colciencias.

Wade, Peter. 1991. "The Language of Race, Place and Nation in Colombia." *America Negra* 2, 41–68.

———. 1993a. *Blackness and Race Mixture: The Dynamics of Racial Identity in Colombia*. Baltimore: Johns Hopkins University Press.

———. 1993b. "El movimiento negro en Colombia." *America Negra* 5, 173–91.

———. 1995. "The Cultural Politics of Blackness in Colombia." *American Ethnologist* 22 no. 2, 341–57.

———. 1996. "Identidad y etnicidad." *Pacífico ¿desarrollo o diversidad? Estado, capital y movimientos sociales en el Pacífico colombiano*, ed. Arturo Escobar and Alvaro Pedrosa, 283–98. Bogotá: CEREC and Ecofondo.

———. 1999. "Working Culture: Making Cultural Identities in Cali, Colombia." *Current Anthropology* 40 no. 4, 449–71.

———. 2002. "Introduction: The Colombian Pacific in Perspective." *Journal of Latin American Anthropology* 7 no. 2.

Wainwright, Joel. 2008. *Decolonizing Development: Colonial Power and the Maya*. London: Blackwell.

Warren, Kay B., and Jean E. Jackson, eds. 2002. *Indigenous Movements, Self-Representation, and the State in Latin America*. Austin: University of Texas Press.

Watts, Michael. 1995. "'A New Deal in Emotions': Theory and Practice and the Crisis of Development." *Power of Development*, ed. Jonathan Crush, 44–62. London: Routledge.

West, Robert C. 1957. *The Pacific Lowlands of Colombia: A Negroid Area of the American Tropics*. Baton Rouge: Louisiana State University Press.

Westwood, Sallie, and Sarah A. Radcliffe. 1993. "Gender, Racism and the Politics of Identities in Latin America." *"Viva": Women and Popular Protest in Latin America*, ed. Sarah A. Radcliffe and Sallie Westwood, 1–29. London: Routledge.

Whitten Jr., Norman. E. 1986. *Black Frontiersmen: Afro-Hispanic Culture of Ecuador and Colombia*. Prospect Heights, Ill.: Waveland Press.

Whitten Jr., Norman E., and Nina S. de Friedemann S. 1974. "La cultura negra del litoral ecuatoriano y colombiano: Un modelo d adaptación étnica." *Revista Colombiana de Antropología* 17, 75–115.

Whitten Jr., Norman E., and Arlene Torres. 1992. "Blackness in the Americas." *Report on the Americas* 25 no. 4, 16–22.

———. 1998. "To Forge the Future in the Fires of the Past: An Interpretive Essay on Racism, Domination, Resistance, and Liberation." *Blackness in Latin America and the Caribbean: Social Dynamics and Cultural Transformations*, ed. Norman E. Whitten Jr. and Arlene Torres, 3–71. Bloomington: Indiana University Press.

Williamson, John. 2002. "What Washington Means by Policy Reform." *Latin American Adjustment: How Much Has Happened?* ed. John Williamson. Institute for International Economics. http://www.iie.com/.

Wolf, Eric. 1982. *Europe and the Peoples without History*. Berkeley: University of California Press.

World Commission on the Environment and Development. 1987. *Our Common Future*. Oxford: Oxford University Press.

Wouters, Mieke. 2001. "Ethnic Rights under Threat: The Black Peasant Movement against Armed Groups' Pressure in the Chocó, Colombia." *Bulletin of Latin American Research* 20 no. 4, 498–519.

Yashar, Deborah J. 1998. "Contesting Citizenship: Indigenous Movements and Democracy in Latin America." *Comparative Politics* 31 no. 1, 23–42.

Zuluaga, Francisco U. 1986. "El Patia: Un caso de producción de una cultura." *La participación del negro en la formación de las sociedades latinoamericanas*, ed. Alexander Cifuentes, 81–96. Bogotá: Colcultura and ICANH.

———. 1994. "Conformación de las sociedades negras del Pacífico." *Historia del Gran Cauca* 13, 243–58.

ACADESAN, (Asociación Campesina del Río San Juan), 40, 44; Law 70 and, 54

ACAPA (Consejo Comunitario de Comunidades Negras del Río Patía Grande, sus Brazos, y la Ensenada), 188

Acción Social (Social Action), 169, 174, 208n12, 209n13; Land Protection project, 188; Law 387 and, 170–71

ACIA (Asociación Campesina Integral del Río Atrato), 37, 38, 39, 40, 44, 45, 49; black land rights and, 39–40, 165; institutional politics and, 53–54; Law 70 and, 54; Women's Section, 138, 139

Aerial fumigation, 166–67, 187; effects of, 23, 157, 166

African oil palm, 23, 60, 158, 167; expansion of cultivation, 168, 169; US interest in, 168

Afro-Colombian/black women; activism of, 28, 131, 137, 149–50, 151; in agro-industrial sector, 132; biodiversity conservation and, 26, 149, 150; black movements and, 30, 131, 136, 137–38; as development category, 151, 152; displacement of, 138; economic development and, 205n2; engagement with development interventions, 149, 150, 152, 153; equality and, 139; Equilibrium Foundation and, 187; ethnic and territorial rights, 147, 148; Gender and Development and, 134; gender concerns of, 133, 139, 140, 141, 206n4; identity of, 135, 136, 137, 142–43, 148, 207n11; Law 70 and, 142; leaders, 137, 138; New constitution and, 139; PCN and, 54; rights of, 134; self-help groups and, 143, 145; social welfare programs and, 62; State policy regarding, 132; *tener* needs of, 133, 134, 138, 151–52; traditional forms of organization, 131, 144, 145; whitening of, 143

Afro-Colombians/blacks, 2, 5, 36, 41, 73; assimilation of, 34, 35; demographic data on, 34, 200n4; elites, 37, 60; in exile, 181; heterogeneity of, 18, 27, 126, 127–28; interethnic relations and, 5, 35, 86–87; invisibility of, 4, 33, 37, 48, 155, 173, 175; isolation of, 34; national coalitions, 43; Law 70 and, 183–84, 196; oral traditions, 43, 146, 205n5; population census of 179; representation in state institutions, 23, 51, 103, 184, 196; sustainable economic practices of, 39, 69; traditional politics and, 179–80

Afro-Colombian Subdivision, Office of Ethnic Affairs, 184

AFRODES (Asociación de Afrocolombianos Desplazados), 175, 181–82, 183, 186

Afro-descendants, 28, 155. *See also* Afro-Colombians/blacks

Afro-Latinos, 155; land rights of, 22

Development (*cont.*)
 history of, 8–10; participatory
  approach, 62; resistance and, 20,
  25, 26, 27, 28, 126,127, 189; socioeconomic, 5, 50, 51, 82; state intervention in, 22–23, 24, 58, 60–61; statistics
  for Pacific region, 162; studies of,
  23–5; traditional knowledge and, 87.
  See also decentralization; development
  plans; PDCN; Plan Pacífico; Proyecto
  BioPacífico; sustainable development
Developmental geopolitics and environmental biopolitics, 77
Developmentalism, 9, 24; failure of,
  63–64
Development plans: black politicians
  and, 124; ethnic concerns as part of,
  27; failure of, 61, 63; national focus
  of, 161–62; social movements and, 7,
  8; state intervention and, 6, 22–23,
  24, 37, 56, 58, 60–61, 91–92, 129;
  sustainable resource management
  and, 159. *See also* Black Women's Networks; development; Plan Pacífico;
  PDCN
DIAR, 39 62
Diocese of Tumaco, 165
Displaced people: AFRODES and,
  181–82, 186; black identity of, 182;
  black movements and, 177; Law 387
  and, 170, 208n12, 209n15; property
  rights of, 170–71; return of, 182, 183;
  state apathy towards, 183; *See also*
  displacement; Social Action
Displacement, 4, 23, 28, 159, 174, 175,
  182–83; coca eradication and, 173, 187;
  collective land titles and, 163, 164;
  figures, 158, 163; invisibility and, 155;
  PCN policy and, 173, 186; Proyecto
  Protección de Territorios y Patrimonio de la Población Desplazada,
  170–71; violence and, 89, 108, 151, 155,
  161, 163, 165, 188; of women, 138
División de Asuntos las Communidades
  Negras (Division of Black Community Affairs), 51, 52, 55, 104, 111, 171
DNP (Departamento Nacional de
  Planeación), 38, 58, 59, 63, 65, 82;
  development plan for black commu-

nities, 182; policies of, 161; redrafting
 Plan Pacífico, 67
Doña Naty, 98, 114, 115–17, 119, 120, 188
Drug cartels, 154
Drug trafficking, 2, 10, 23, 61, 87, 111,
 150, 165; campaign against, 156, 157

Ecofondo, 73, 146
Ecological data, 78
Ecological research, 85, 88
Ecological sustainability: development
 and, 62; state and, 27, 60. *See also*
 sustainable development
Ecological zones, 90
Ecological zoning, 65, 66, 78–79. *See also*
 territorial zoning
Economic opening. *See apertura
 económica*
Economic reforms, 2, 10, 12. *See also
 apertura económica*
Ecopetrol pipeline, 15, 16
Ecos del Pacífico, 124
Education, 11, 44; of displaced blacks,
 182; ethno, 51, 67, 112, 113, 195; Law
 70 and, 50–51, 161, 195; Plan Pacífico
 and, 65; women and, 136, 145
ELN (Ejército de Liberación Nacional),
 11, 156
El proceso, 103, 114, 120; PCN and, 124,
 128; women's contribution, 137–38,
 140
Embera indians, 36, 38, 62
Empty lands (*tierras baldías*), 37, 197n2;
 black collective rights over, 2, 5, 47,
 192; illegal logging in, 80
Environment: World Commission on
 Environment and Development,
 13; United Nations Conference on
 Environment and Development (Rio
 de Janerio), 13, 39; United Nations
 Conference on the Human Environ-
 ment, 13. *See also* conservation
Environmental Law (Law 99 of 1993),
 70, 78
Environmental NGOS: IIAP and, 71,
 Law 70 and, 193; research institutes,
 70. *See* Equilibrium Foundation;
 FES; Fundacion Habla/Scrib; Fundación Herencia Verde

Mejía, Claudia, 169–70, 171, 188

Mena, Elizabeth, 52–53

Mena, Zulia, 4, 12, 69, 72, 137, 179, 181; election to Chamber of Representatives, 52, 54

Mestizaje, 32, 33, 34

Mestizos, 14, 33, 60

Migration. *See* displacement

Mina, María del Rosario, 151

Mining, 119, 120; concessions, 60; Law 70 and, 194

Ministery of Environment, Housing, and Territorial Development, 162

Ministery of the Environment (Ministerío de Medio Ambiente), 14, 70, 72; criticism of, 88; defense of Ecopetrol pipeline, 16; draft statutes for IIAP, 71, 72, 73, 88, 159; Law 70 and, 82

Mobilization of black people, 6, 36, 45, 56; AT 55 and, 5, 46–48, 54, 137; black politician's role in, 44; Black Women's Network and, 140–41, 142; catholic church and, 33, 37, 44; constitutional reforms and, 44; difficulties of, 102, 120–21; ethno-cultural strategy for, 52, 68; gender and, 121; OCN and, 43, 48; PCN and, 17, 121; role of NGOS in, 44; socio-cultural issues and, 43; state and, 18–19, 20, 48. *See also* OCN; PCN

Modernity, 8; expansion of, 111; hybrid cultures and, 24, 25

Modernization, 2, 129; agricultural, 11, 41, 60; black culture and, 4–5; economic, 16; resistance to, 17, 25, 59, 128; state and, 188. *See also* development; globalization; Plan Pacífico

Moreno, Camila, 109, 169–70

Moreno, Saturnino, 3–4

Mosquera, Juan de Dios, 42, 44–45, 185

Movimiento Nacional Cimarrón. *See* Cimarrón

Movimiento 19 de Abril (April 19 Movement), 11, 201n9

Moya, Margarita, 137

Muelas, Lorenzo, 45, 47

Multiculturalism, 4, 13, 20

Multiethnicity, 46, 127

Multilateral funding, 156, 157, 181, 182, 207n5; Plan Pacífico and, 58, 66–67; reforms and, 21

Murillo, Luis Gilberto, 181

Nariño, 5, 43, 54, 60, 62, 109, 120, 157, 165, 187

National Conference of Black Communities: First (Tumaco), 47–48; Second (Bogotá), 50; Third (Puerto Tejada), 51, 101

National Constituent Assembly, 12, 13, 47; debates over black rights, 45, 46; draft of new constitution, 11–12; election of, 41, 44–45, 198n9; ethnic and cultural rights and, 41–42; indigenous representatives in, 12, 13, 45; members of, 201n9; Subcommission on Equality and Ethnic Rights, 45

National Coordinator for Black Communities, 45

National Council of Palenques (Consejo Nacional de Palenques), 113

National Environmental Network (SINA), 70, 71

National Front, 10, 11

Nationalism, 9, 25, 32

Nationalist ideologies, 32–33

National parks and nature reserves, 166–67

Nation state; 18, 20, 33, 90, 102, 173

Natividad Urrutia. *See* Doña Naty

Natural resources, 2, 6; exploitation of, 87; management of, 129; protection under Law 70, 194. *See also* resource management; PMRN

Nature, 85; regimes of production, 19

Negritude: ideology of, 32; movements, 42

Neoliberal globalization, 6, 10, 21, 22, 122, 129, 149

Network for Advocacy in Solidarity with Afro-Colombian Grassroots Communities, 209n13

New social movements, 17–18, 25–26, 127; hybrid model of resistance and, 25–26

Nongovernmental Organizations (NGOS), 8, 62, 63; black mobilization and, 44; environmental and social concerns of, 21; participation in IIAP workshops, 73; PCN and, 124. *See also* ACIA; FES; Fundación Habla/Scribe; Fundación Herencia Verde; Fundación Inguede; Fundación Natura

Ng'weno, Bettina, 64

OBAPO (Organización de Barrios Populares y Comunidades Negras de Chocó), 38, 44; black mobilization in Chocó, 38, 43; grassroots activism and, 52–53; women activists of, 139

Observatory of Racial Discrimination, 162, 185

OCN (Organización de Comunidades Negras), 43–44, 45, 54; collective ethnic identity and, 4–5; emphasis on ethno-cultural development, 48; vision of black rights, 4–5, 7, 44, 49. *See also* PCN

Oral history, 43; Patía project, 82, 83–85

Oral traditions, 43, 146, 205n5. *See also coplas*

Ordenamiento territorial. See territorial zoning

OREWA (Organización Regional Emberá-Waunana del Chocó), 38, 72, 86

Organic nature, 19

Ortiz, Willington, 179–80, 202n17, 209n16

Palacio, Andres, 184

Palacios, Marco, 11

Palenqueros, 114

Palenques, 34, 110, 204n3; regional, 109, 113

Paramilitary forces, 2, 10, 23, 89, 90, 111, 120, 150, 154, 208n9; agribusiness interests and, 167; demobilization of, 168, 174; guerillas and, 163, 164, 165; state military and, 207n4, 208n7. *See also* AUC

Pardo, Maurico, 128

Parks and reserves, 90

Participatory research, 83, 84, 85

Pastrana, Andres, 156

Patía: community councils of, 83, 188; Defense Council, 83; oral history project, 82, 83–5; river, 80

PCN (Proceso de Comunidades Negras), 7, 22, 26, 52, 54–6, 101–2, 105, 106, 118, 119, 125, 155; African diasporas and, 185–86; autonomy and, 20, 55, 89, 101, 102–103, 128; black identity and, 20, 55, 127; black mobilization and, 17, 102, 121; black movements and, 27, 101, 173; Black Women's Networks and, 140–41; Cimarrón and, 185; community councils and, 113; emphasis on collective decision making, 105, 106, 114; emphasis on territory, 20, 55, 185; *espacios propios* and, 101; ethnic rights and, 59, 185; ethno-cultural strategy, 17, 21, 27, 30, 55–56, 103, 108–9, 123, 125, 127, 129, 140, 150; gender concerns and, 139–40, 141, 152; institutional politics and, 54–55, 179–80; new strategies of, 175–76, 177, 178; opposition to black politicians, 107, 108; opposition to Cali-Buenaventura pipeline, 15, 16; opposition to development plans, 108–9; 128; *palenques* and, 110; politics of resistance and, 101, 127, 129; relations with NGOs, 123, 124; state institutions and, 20–21, 123, 128, 129; transnational activism and, 123, 175–76; urban blacks and, 108; vision of PDCN, 112; women's role in, 31, 54, 131

PDCN (Plan de Desarrollo de Comunidades Negras), 27, 108, 109, 110, 160; as alternative model of development, 121; Chocoanización of, 111; as cultural and political manifesto, 112; state development processes and, 125

Peace and Justice Law (Law 975 of 2005), 185

Pedrosa, Alvaro, 106

Perico Negro, 85–89, 113, 159

PLADEICOP (Plan de Desarrollo Integral para la Región Pacífico), 61–62, 132

Plan Colombia, 156–57, 207n2

collective land titles for, 54; mobilization of, 43, 44, 45; role in AT 55, 54; subsistence activities of, 18, 54, 112–13
Robledo, Onni, 52
Rodriguez, Sylveria, 130, 134, 152
Rojas, Jeannette, 132, 188
Rojas Birry, Francisco, 45, 47
Roldán, Roque, 35
Romana, Geilerre, 181, 182, 183
Rosero, Carlos, 15, 16, 44, 103, 124, 176, 179, 180, 181, 183, 184; on black and indigenous movements, 17; on black diversity, 127–28; on institutional politics, 54–55; international networking and, 123, 185–86; on new resource management laws, 168–69; new strategy for black movements, 177–78; on PCN's vision and strategy, 125–26; on PDCN, 112, 125; on racial discrimination, 185; on state development plans, 109
RSS (Red de Solidaridad Social), 65, 82, 109–10, 138, 159
Ruiz, Juan Pablo, 76
Rural communities: displacement of, 164, 173; environmental conservation and, 14; marginalization of, 37, 40–41; mobilization of, 43, 44, 45
Rural Development Law (Law 1152), 171, 172

Salazar, Fernando, 78
Samper, Ernesto, 65, 67, 109, 154
San Andrés and Providencia islands, 36, 162
Sanquianga river, 62, 79, 80, 96
Satinga river, 62, 81
Savings and Loan Cooperative of Women Producers of Guapi (Coop-Mujeres), 99, 130, 131, 132, 133, 134, 135, 136, 187
Scott, Joan, 126
Ser (right to be), 5, 134, 151–52, 153, 182. *See also* black identity
Ser Mujer (Ser Mujer Cooperativa de Ahorro y Crédito), 132–33, 136
Shrimp farming, 23, 132

Siglo XXI (Fundación Litoral Siglo XXI), 43, 45
Sistema Nacional Ambiental (SINA), 70, 71
Slaves, 18, 34, 36, 69, 110, 204n3
Social movements: capitalist globalization and, 1, 17, 189; cultural politics and, 8, 19, 20; indigenous, 13, 14, 17; new, 17–18, 25–26, 127; state intervention and, 150; Third World, 1, 17–18, 24, 25. *See also* black social movements
Social welfare programs, 61, 62, 65
Spivak, Gayatri, 30, 127
State: Afro-Colombian rights and, 6, 16, 49; biodiversity conservation and, 92, 129; biopolitics/biopower and, 92, 93; black ethno-cultural politics and, 128; black social movements and, 20, 128; developmental interventions by, 22–23, 24, 58, 60–61; ethnic rights and, 59, 60; Gramscian formulation of, 93; PCN's engagement with, 20–21, 128, 129; policy regarding women, 132; protest against, 11, 13; territorial autonomy and, 89; territorial zoning and, 91, 92, 158
State Forest Reserves, 37
Structural adjustment policy, 21, 64, 65
Subcommission on Equality and Ethnic Rights, 45–47
Subsistence activities, 5, 6, 18, 36, 39, 60, 80, 81, 112–13, 114, 117; legal recognition under Law 70, 51, 194
Sustainable development, 13, 73; black livelihood strategies and, 39; free trade and, 21; international interest in, 111; land claims and, 39; local participation in, 14; social component of, 39, 88; state and, 129; women's networks and, 21, 26. *See also* Plan Pacífico; Proyecto BioPacífico

Tagua (vegetal ivory), 36, 60
Taussig, Michael, 35
Technonature, 19
Tener (to have), 133, 134, 138, 151–52, 153

KIRAN ASHER IS ASSOCIATE PROFESSOR OF

INTERNATIONAL DEVELOPMENT AND SOCIAL CHANGE,

AND WOMEN'S STUDIES AT CLARK UNIVERSITY.

Library of Congress Cataloging-in-Publication Data
Asher, Kiran.
Black and green : Afro-Colombians, development, and nature in
the Pacific lowlands / Kiran Asher.
p. cm.
Includes bibliographical references and index.
ISBN 978-0-8223-4487-2 (cloth : alk. paper)
ISBN 978-0-8223-4483-4 (pbk. : alk. paper)
1. Blacks—Colombia—Pacific Coast—Social conditions.
2. Blacks—Civil rights—Colombia—Pacific Coast. 3. Blacks—
Race identity—Colombia—Pacific Coast. 4. Blacks—Colombia—
Pacific Coast—History. 5. Civil rights movements—Colombia.
6. Pacific Coast (Colombia)—Civilization—African influences.
I. Title.
F2299.B55A84 2009
305.896'08615—dc22   2009005700